Channel-Based Single-Queue SBC
Process Algebra
For Systems Modeling

-- General Systems Theory 2.0 at Work --

William S. Chao

Structure-Behavior Coalescence

Systems Architecture = Systems Structure + Systems Behavior

CONTENTS

CONTENTS 5
PREFACE 9
ABOUT THE AUTHOR 11
PART I: BASIC CONCEPTS 13
    Chapter 1: Introduction to Systems Modeling 15
        1-1 Physical and Virtual Systems 15
        1-2 Boundary and Environment of a System 17
        1-3 Systems Modeling 1.0 19
        1-4 Systems Modeling 2.0 21
    Chapter 2: Introduction to Process Algebra 23
        2-1 What is Process Algebra? 23
        2-2 Examples of Process Algebras 23
        2-3 Generalized SBC Process Algebra 23
        2-4 Specialized SBC Process Algebras 24
    Chapter 3: Mathematics of SBC Process Algebra 25
        3-1 Sequentialization of Prefixes 25
        3-2 Summation of Processes 25
        3-3 Parallel Composition of Processes 25
        3-4 Recursive Definition of a Process 26
        3-5 Replication of a Process 26
        3-6 Conditional Definition of a Process 26
        3-7 Null Process 26
PART II: CHANNEL-BASED SINGLE-QUEUE SBC PROCESS ALGEBRA 29
    Chapter 4: Channel-Based Value-Passing Interactions 31
        4-1 Channels 31
        4-2 Channel-Based Interactions 32
        4-3 Formal Description of a Channel-Based Communication Port 35
        4-4 Formal Description of a Channel-Based Action 36
        4-5 Formal Description of a Channel-Based Interaction 37
    Chapter 5: The Structure-Behavior Coalescence Approach 39
        5-1 Structure-Behavior Coalescence Means to Integrate the Systems Structure and Systems Behavior 39
        5-2 Interactions among Components and the External Environment to Draw Forth the Systems Behavior 40

5-3 Core Theme of Structure-Behavior Coalescence..................................41

Chapter 6: Language Constructs of Channel-Based Single-Queue SBC Process Algebra 43

    6-1 Entity Set and Entity Name 43

    6-2 Backus-Naur Form of Channel-Based Single-Queue SBC Processes...44

        6-2-1 Recursion of Summation of One or More Interaction Flow Diagrams Defines the Channel-Based Single-Queue SBC Process of a System 45

        6-2-2 An Interaction Flow Diagram Consists of a Type_1 Expression, Followed by Zero or More Type_1_Or_2 Expressions 46

        6-2-3 Type_1_Or_2 Expression is either Type_1 or Type_2 46

        6-2-4 Type_1 Expression is either a Called Action with Optional Variable Settings or a Conditional Expression of Called Action with Optional Variable Settings 47

        6-2-5 Type_2 Expression is either an Interaction with Optional Variable Settings or a Conditional Expression of Interaction with Optional Variable Settings 47

Chapter 7: Transitional Semantics of Channel-Based Single-Queue SBC Process Algebra 49

    7-1 Transitional Semantics..49

    7-2 Rule of Prefix..50

    7-3 Rule of Summation 51

    7-4 Rule of Recursion 51

    7-5 Rule of Constants..52

PART III: CASES STUDY..53

    Chapter 8: Channel-Based Single-Queue SBC Process of the Robot System 55

        8-1 BNF Tree of the Robot System...55

        8-2 Prefixes of the Robot System..57

        8-3 Process of the Robot System 59

    Chapter 9: Channel-Based Single-Queue SBC Process of the Multi-Tier Personal Data System 61

        9-1 BNF Tree of the Multi-Tier Personal Data System 65

        9-2 Prefixes of the Multi-Tier Personal Data System 68

        9-3 Process of the Multi-Tier Personal Data System 75

    Chapter 10: Channel-Based Single-Queue SBC Process of the Mathematical Calculation System..77

        10-1 BNF Tree of the Mathematical Calculation System 80

10-2 Prefixes of the Mathematical Calculation System 83

10-3 Process of the Mathematical Calculation System 88

Chapter 11: Channel-Based Single-Queue SBC Process of the Web Service Arithmetic System 91

11-1 BNF Tree of the Web Service Arithmetic System 94

11-2 Prefixes of the Web Service Arithmetic System 95

11-3 Process of the Web Service Arithmetic System 100

Chapter 12: Channel-Based Single-Queue SBC Process of the Web Service Extranet System 103

12-1 BNF Tree of the Web Service Extranet System 106

12-2 Prefixes of the Web Service Extranet System 108

12-3 Process of the Web Service Extranet System 113

Chapter 13: Channel-Based Single-Queue SBC Process of the Convenience Store's Get 2nd 50% off Sales Promotion System 115

13-1 BNF Tree of the Convenience Store's Get 2nd 50% off Sales Promotion System 115

13-2 Prefixes of the Convenience Store's Get 2nd 50% off Sales Promotion System 118

13-3 Process of the Convenience Store's Get 2nd 50% off Sales Promotion System 120

Chapter 14: Channel-Based Single-Queue SBC Process of the Department Store's Car Sweepstakes Sales Promotion System 121

14-1 BNF Tree of the Department Store's Car Sweepstakes Sales Promotion System 121

14-2 Prefixes of the Department Store's Car Sweepstakes Sales Promotion System 124

14-3 Process of the Department Store's Car Sweepstakes Sales Promotion System 129

Chapter 15: Channel-Based Single-Queue SBC Process of the Female Mantis 131

15-1 BNF Tree of the Female Mantis 131

15-2 Prefixes of the Female Mantis 133

15-3 Process of the Female Mantis 136

Chapter 16: Channel-Based Single-Queue SBC Process of the Human Body 139

16-1 BNF Tree of the Human Body 139

16-2 Prefixes of the Human Body 142

16-3 Process of the Human Body 147

Chapter 17: Channel-Based Single-Queue SBC Process of the Disaster 149
    17-1 BNF Tree of the Disaster 149
    17-2 Prefixes of the Disaster 152
    17-3 Process of the Disaster 157

Chapter 18: Channel-Based Single-Queue SBC Process of the Bicycle 159
    18-1 BNF Tree of the Bicycle 159
    18-2 Prefixes of the Bicycle 162
    18-3 Process of the Bicycle 166

Chapter 19: Channel-Based Single-Queue SBC Process of the Restaurant 169
    19-1 BNF Tree of the Restaurant 169
    19-2 Prefixes of the Restaurant 172
    19-3 Process of the Restaurant 178

Chapter 20: Channel-Based Single-Queue SBC Process of the Car 181
    20-1 BNF Tree of the Car 181
    20-2 Prefixes of the Car 184
    20-3 Process of the Car 188

APPENDIX A: Language Constructs of Channel-Based Single-Queue SBC Process Algebra 189

APPENDIX B: Transitional Semantics of Channel-Based Single-Queue SBC Process Algebra 191

BIBLIOGRAPHY 193

INDEX 197

PREFACE

The need for systems modeling arises because any real-life system is inherently complicated. It is impossible to comprehend fully the intricate interaction of any system of the real world with its environment, or to define all its components and each of its details. Systems modeling or system modeling is an artifact created by humans to define what a system is.

Process algebras are a diverse family of related approaches to the study of concurrent systems. Their tools are algebraic languages for the high-level description of interactions, communications and synchronizations between a collection of independent agents or processes. Process algebras also provide algebraic laws that allow process descriptions to be manipulated and analyzed, and permit formal reasoning about equivalences and observation congruence among processes. Accordingly, process algebra provides a perfect method for system modeling.

Channel-based single-queue SBC process algebra (C-S-SBC-PA) is one of the six specialized SBC process algebras. In this book, we use C-S-SBC-PA to achieve the robust systems modeling of a system. To see is to believe. Therefore, many examples are presented to help the reader fully understand the use of C-S-SBC-PA.

ABOUT THE AUTHOR

Dr. William S. Chao is the CEO & founder of SBC Architecture International®. SBC (Structure-Behavior Coalescence) architecture is a systems architecture which demands the integration of systems structure and systems behavior of a system. SBC architecture applies to hardware architecture, software architecture, enterprise architecture, knowledge architecture and thinking architecture. The core theme of SBC architecture is: "Architecture = Structure -->> Behavior."

William S. Chao received his bachelor degree (1976) in telecommunication engineering and master degree (1981) in information engineering, both from the National Chiao-Tung University, Taiwan. From 1976 till 1983, he worked as an engineer at Chung-Hwa Telecommunication Company, Taiwan.

William S. Chao received his master degree (1985) in information science and Ph.D. degree (1988) in information science, both from the University of Alabama at Birmingham, USA. From 1988 till 1991, he worked as a computer scientist at GE Research and Development Center, Schenectady, New York, USA.

Dr. William S. Chao has been teaching at National Sun Yat-Sen University, Taiwan since 1992 and now serves as the president of Association of Enterprise Architects, Taiwan Chapter. His research covers: systems architecture, hardware architecture, software architecture, enterprise architecture, knowledge architecture and thinking architecture.

PART I: BASIC CONCEPTS

14

Chapter 1: Introduction to Systems Modeling

All things that amaze us as something independent are essentially parts of a system. We usually call the parts of a system its components. Every system is something the whole. Systems emphasize the holistic vision.

The need for systems modeling arises because any real-life system is inherently complicated. It is impossible to comprehend fully the intricate interaction of any system of the real world with its environment, or to define all its components and each of its details. Systems modeling or system modeling is an artifact created by humans to define what a system is.

In this chapter, we first introduce physical and conceptual systems. A physical system exists in the physical, concrete, or real world. A conceptual system exists in the conceptual, abstract, or virtual world. We then introduce the boundary of a system. A system has a boundary. The system itself is inside the boundary and the environment is outside the boundary. After these introductions, we shall thereafter discuss the systems modeling 1.0 as well as the systems modeling 2.0.

1-1 Physical and Virtual Systems

In general, systems are divided into two categories: 1) physical systems and 2) virtual systems.

A physical system exists in the physical world [Acko68]. A physical system is also called a concrete or real system. For example, a *telephone* composed of *microphone, earphone and keypad*, shown in Figure 1-1, is a physical, concrete, or real system.

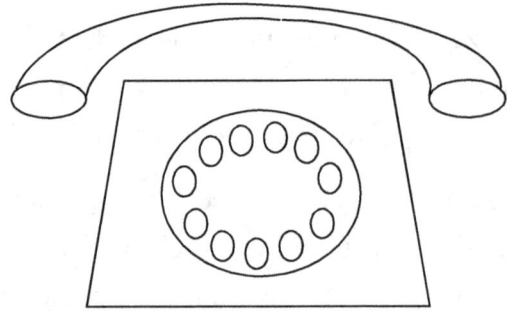

Figure 1-1   A Telephone is a Physical System

As a second example, a *stool* composed of *seat* and *legs*, shown in Figure 1-2, is a physical, concrete, or real system.

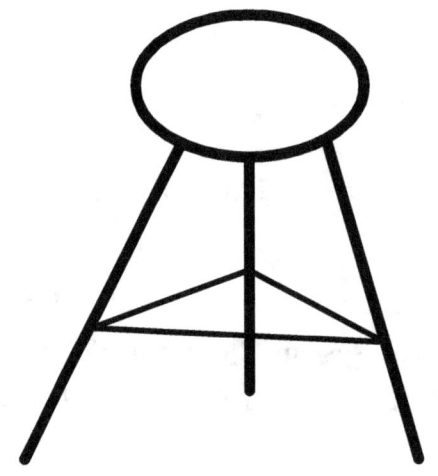

Figure 1-2   A Stool is a Physical System

A virtual system is a system that is composed of non-physical components, i.e., ideas, thoughts, or notions. A virtual system exists in the virtual, abstract, or notional world. For example, the "*Snow White and the Seven Dwarfs*" fairy tale composed of "*Snow White*" and "*Seven Dwarfs*," shown in Figure 1-3, is a virtual, abstract, or notional system.

Figure 1-3 *Snow White and the Seven Dwarfs* is a Virtual System

As a second example, the "*Multi-Tier Personal Data System*" software composed of *MTPDS_GUI*, *Age_Logic*, *Overweight_Logic* and *Personal_Database*, shown in Figure 1-4, is a virtual, abstract, or notional system.

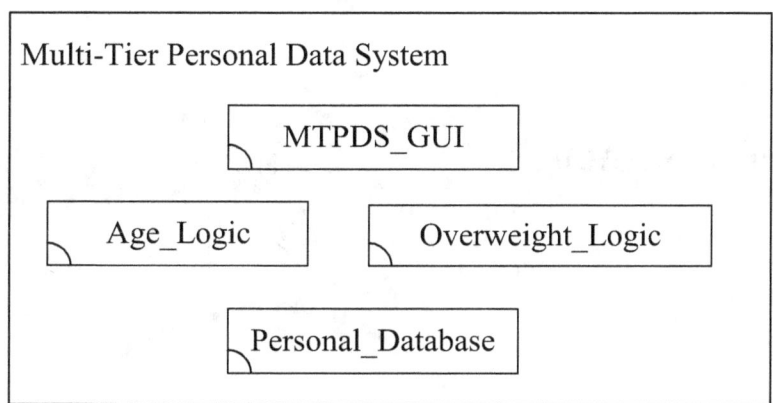

Figure 1-4 *Multi-Tier Personal Data System* is a Virtual System

1-2 Boundary and Environment of a System

We scope a system by describing its boundary as shown in Figure 1-5. All components of the system are inside the boundary while the environment is outside the boundary.

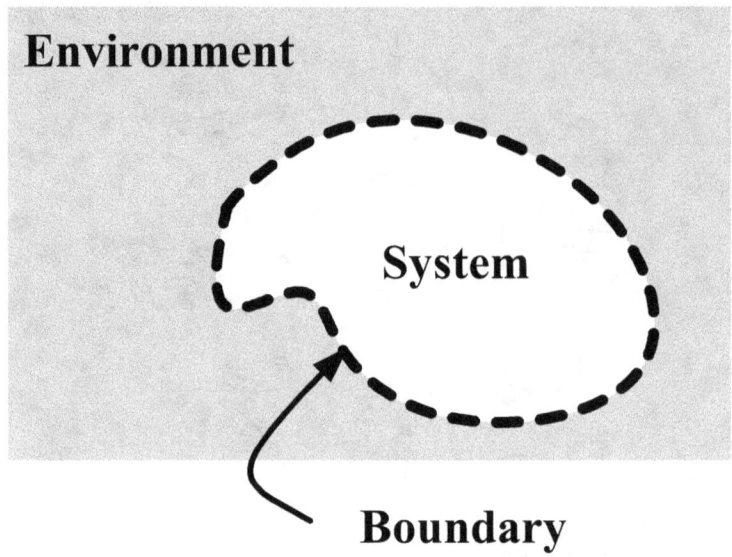

Figure 1-5   Boundary and Environment of a System

The environment is also known as the surroundings. A system may or may not interrelate with the environment. An open system interrelates with the environment through the exchange of matter, energy, data, information, or message as shown in Figure 1-6.

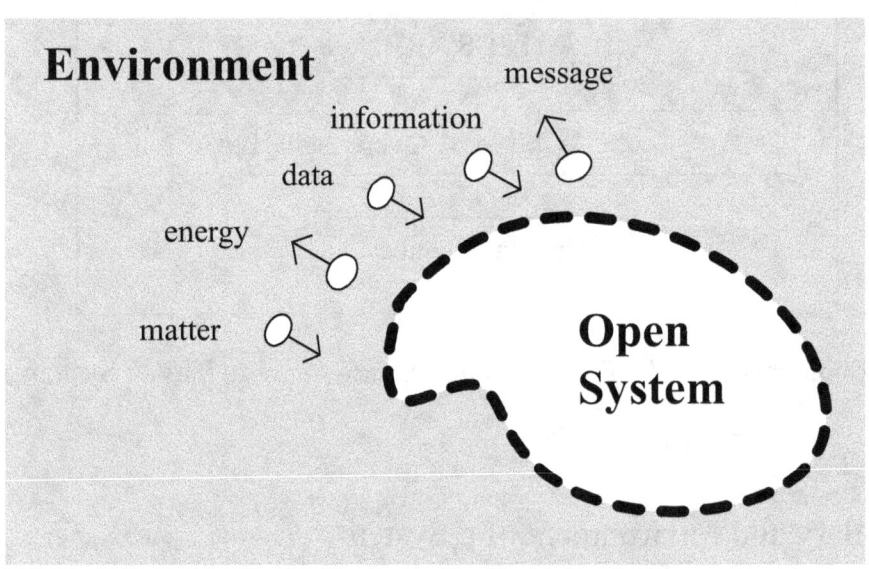

Figure 1-6   Open System Interrelates with the Environment

An isolated system does not interrelate with the environment at all. There is no exchange of matter, energy, data, information, or message between the isolated system and the environment as shown in Figure 1-7.

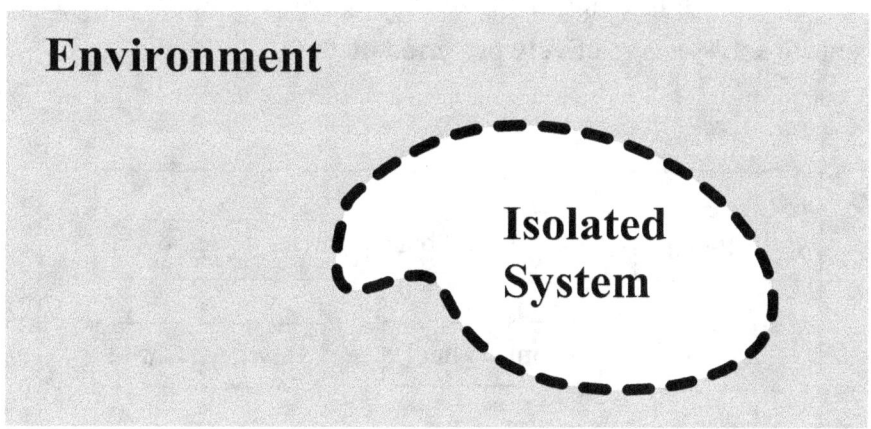

Figure 1-7   Isolated System Does Not Interrelate with the Environment

1-3 Systems Modeling 1.0

Systems modeling 1.0 defines a system, in Figure 1-8, hopefully to be an integrated whole, embodied in its assembled components, their interrelationships with each other and the environment [Chec99, Frie11, Ghar11, Mead08].

> A system, hopefully is an integrated whole,
> embodied in its assembled components,
> their interrelationships with each other and the environment.

Figure 1-8   Systems Modeling 1.0 Defining a System

Components are sometimes labeled as parts, entities, objects, building blocks and non-aggregated systems [Chao14a, Chao14b, Chao14c]. Interrelated components make a system not only a whole but also hopefully an integrated whole.

A system defined by systems modeling 1.0 has the following characteristics: 1) hopefully, it is an integrated whole; 2) it is embodied in its assembled components; 3) components are interrelated with each other and the environment; and 4) it uses structural decomposition [Chao14b, Ghar11] rather than functional decomposition [Scho10].

The structural decomposition method is to decompose a system into a number of components, as shown in Figure 1-9. Breaking down a large problem into a number of components to solve, is a relatively preferred method.

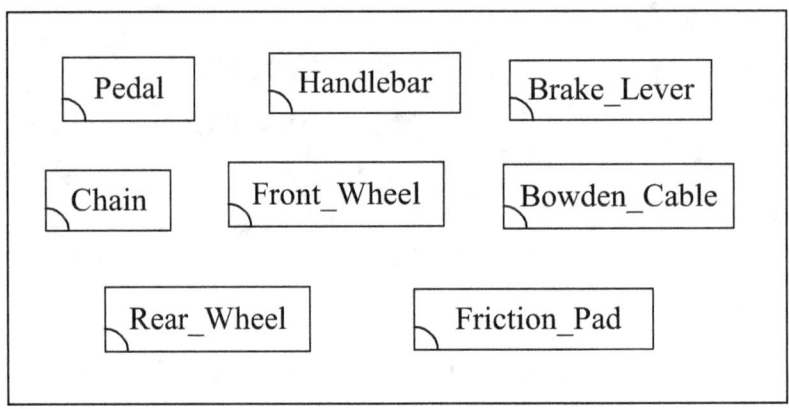

Figure 1-9 Structural Decomposition Method

The functional decomposition method is to decompose a system into a number of functions, as shown in Figure 1-10. Breaking down a large problem into a number of functions to solve, is a relatively non-preferred method.

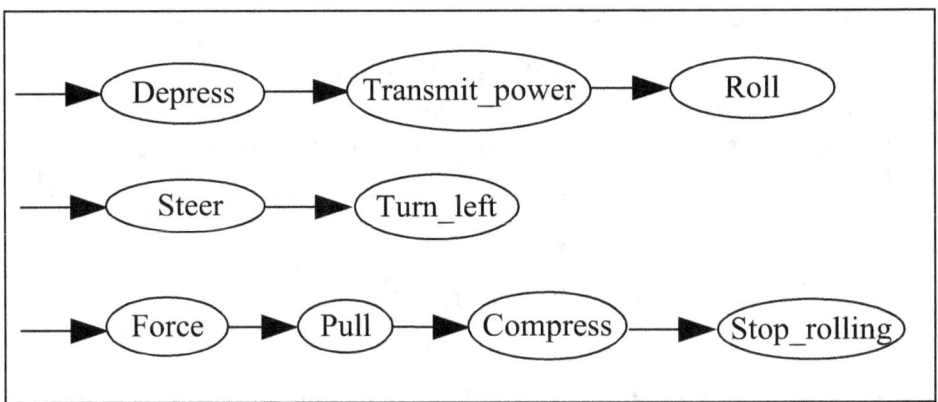

Figure 1-10 Functional Decomposition Method

1-4 Systems Modeling 2.0

Systems structure and systems behavior are the two most significant views of a system. In order to achieve a truly integrated whole of a system, we first need to integrate the systems structure and systems behavior together. In other words, integration of systems structure and systems behavior results in the integration of a whole system. Since systems modeling 1.0 does not define the integration of systems structure and systems behavior, very likely it only hopes and will never be able to really form an integrated whole of a system. In this situation, systems modeling 1.0 is powerless in defining a system appropriately.

Structure-behavior coalescence (SBC) provides an elegant way to integrate the systems structure and systems behavior of a system. A system is therefore redefined, by systems modeling 2.0, truly to be an integrated whole, through structure-behavior coalescence, embodied in its assembled components, their interactions with each other and the environment as shown in Figure 1-11.

A system,
through the SBC approach,
truly is an integrated whole,
embodied in its assembled components,
their interactions with each other and the environment.

Figure 1-11   Systems Modeling 2.0 Defining a System

Systems modeling or system modeling 2.0 formally defines the essence of a system and its details at the same time. Since systems modeling 2.0 demands the integration of systems structure and systems behavior, definitely it is able to form an integrated whole of a system. In this situation, systems modeling 2.0 is fully capable of defining a system.

A system defined by systems modeling 2.0 has the following characteristics: 1) it emphasizes the system's structure-behavior coalescence; 2) it is a truly integrated whole; 3) it is embodied in its assembled components; 4) components are interacting (or handshaking) with each other and the environment; and 5) it uses structural decomposition [Chao14a, Chao14b, Chao14c, Ghar11] rather than functional decomposition [Scho10].

Chapter 2: Introduction to Process Algebra

In this chapter, we first discuss what process algebra is. Then we show several examples of process algebras. After that we elaborate on the generalized SBC process algebra and six specialized SBC process algebras.

2-1 What is Process Algebra?

Process algebras are a diverse family of related approaches to the study of concurrent systems [Berg87, Chao15d, Hoar85, Miln89, Miln99]. Their tools are algebraic languages for the high-level description of interactions, communications and synchronizations between a collection of independent agents or processes.

Process algebras also provide algebraic laws that allow process descriptions to be manipulated and analyzed, and permit formal reasoning about equivalences and observation congruence among processes.

2-2 Examples of Process Algebras

There are several leading algebraic approaches to modeling concurrent systems.

Communicating Sequential Processes (CSP) [Hoar85] was first described in a 1978 paper by C. A. R. Hoare.

Arthur John Robin Gorell Milner introduced the Calculus of Communicating Systems (CCS) [Miln89, Miln99] around 1980.

Algebra of Communicating Processes (ACP) [Berg87] was initially developed by Jan Bergstra and Jan Willem Klop in 1982.

2-3 Generalized SBC Process Algebra

Generalized SBC process algebra (G-SBC-PA) evolved from CCS (Calculus of Communicating Systems) [Miln89, Miln99].

CCS is a general process algebra language for the study of communication and concurrency. Like CCS, generalized SBC process algebra is also a general process algebra language for the study of communication and concurrency.

2-4 Specialized SBC Process Algebras

Channel-based single-queue SBC process algebra (C-S-SBC-PA) [Chao15d, Chao15e, Chao15g], channel-based multi-queue SBC process algebra (C-M-SBC-PA) [Chao15d, Chao15f, Chao15h], channel-based infinite-queue SBC process algebra (C-I-SBC-PA) [Chao15b, Chao15c, Chao15d], operation-based single-queue SBC process algebra (O-S-SBC-PA), operation-based multi-queue SBC process algebra (O-M-SBC-PA) [Chao15d, Chao15f, Chao15h] and operation-based infinite-queue SBC process algebra (O-I-SBC-PA) [Chao15b, Chao15c, Chao15d] are the six specialized SBC process algebras.

Channel-based single-queue SBC process algebra, channel-based multi-queue SBC process algebra, channel-based infinite-queue SBC process algebra, operation-based single-queue SBC process algebra, operation-based multi-queue SBC process algebra and operation-based infinite-queue SBC process algebra all evolved from CCS (Calculus of Communicating Systems) [Miln89, Miln99].

CCS is a general process algebra language for the study of concurrent systems. Unlike CCS, six specialized SBC process algebras are only applicable to systems modeling [Burd10, Maie09, Chao16b, Chao16c, Chao16d, Chao16e, Chao16f, Chao16g, Craw15, Dam06, O'Rou03, Putm00, Rayn09, Roza11, Toga08].

Chapter 3: Mathematics of SBC Process Algebra

To give the SBC process a mathematical definition, we need a means to form new processes from old ones. The basic operators, always present in some form or other, allow sequential composition of processes or summation of processes or parallel composition of processes or recursive definition of a process or replication of processes or conditional definition of a process or null process.

3-1 Sequentialization of Prefixes

Sometimes prefixes must be temporally ordered. For example, it might be desirable to specify algorithms such as: execute the "t" prefix first and then execute the "P" process later. Sequentialization of prefixes can be used for such purposes.

Sequentialization of prefixes, usually written as the $t \bullet P$ process, indicates that it will perform the "t" prefix first and continue as the "P" process.

3-2 Summation of Processes

The binary operator "+", summation, combines two process expressions as alternatives.

For example, the process $P+Q$ can proceed non-deterministically either as the process P or the process Q; as soon as one performs its first action/interaction the other is discarded.

3-3 Parallel Composition of Processes

Parallel composition of two processes P and Q, usually written $P \| Q$, is the key primitive distinguishing the process algebras from sequential models of process executions.

Parallel composition allows the executions in P and Q to proceed simultaneously and independently.

3-4 Recursive Definition of a Process

The operators presented so far describe only finite action/interaction and are consequently insufficient for full computability, which includes non-terminating behavior. Recursion is the operator that allows finite descriptions of infinite behavior.

For example, **fix**($X=E$) can be understood as abbreviating the recursive definition of an infinite behavior denoted by the "X" process variable.

3-5 Replication of a Process

Replication is the other operator that allows finite descriptions of infinite behavior of a process.

For example, replication $!P$ can be understood as abbreviating the parallel composition of a countably infinite number of P processes.

3-6 Conditional Definition of a Process

A process can be defined by a one-or-more-armed conditional expression. For example, the process (**if** $cond_1$ **then** P_1)+(**if** $cond_2$ **then** P_2)…+(**if** $cond_j$ **then** P_j) will proceed as the process P_1 if the "$cond_1$" value is true, or proceed as the process P_2 if the "$cond_2$" value is true,…, or proceed as the process P_j if the "$cond_j$" value is true.

3-7 Null Process

Process algebras generally also include a null process, denoted as *STOP*, which has no interaction points. It is utterly inactive and its sole purpose is to act as the inductive anchor on top of which more interesting processes can be generated.

The process "*STOP*●P_1" (i.e. sequential composition of processes *STOP* and P_1) equals to the process "*STOP*", as shown in Figure 3-1.

$$STOP \bullet P_1 = STOP$$

Figure 3-1   Characteristics of Null Process (I)

The process "P_2+STOP" (i.e. summation of processes P_2 and $STOP$) equals to the process "$STOP+P_2$" (i.e. summation of processes $STOP$ and P_2) which equals to the process "P_2", as shown in Figure 3-2.

$$P_2 + STOP = STOP + P_2 = P_2$$

Figure 3-2   Characteristics of Null Process (II)

The process "$P_3\|STOP$" (i.e. parallel composition of processes P_3 and $STOP$) equals to the process "$STOP\|P_3$" (i.e. parallel composition of processes $STOP$ and P_3) which equals to the process "P_3", as shown in Figure 3-3.

$$P_3 \| STOP = STOP \| P_3 = P_3$$

Figure 3-3   Characteristics of Null Process (III)

PART II: CHANNEL-BASED SINGLE-QUEUE SBC PROCESS ALGEBRA

Chapter 4: Channel-Based Value-Passing Interactions

In this chapter, we first introduce channels and channel-based interactions. We then introduce the formal description of a channel-based communication port, a channel-based action and a channel-based interaction.

4-1 Channels

Channels are a model for agent communication. An agent may provide many channels, as shown in Figure 4-1.

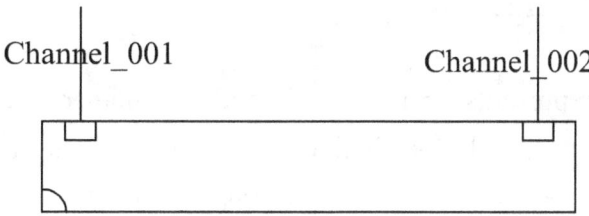

Figure 4-1. A An Agent May Provide Many Channels

A channel may contain several input parameters (e.g. i_1, i_2) and output parameters (e.g. o_1, o_2), as shown in Figure 4-2.

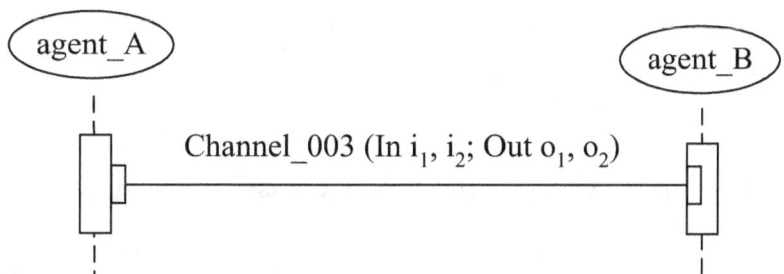

Figure 4-2. A Channel Contains Several Input/Output Parameters

A channel formula is used to completely describe a channel. A channel formula includes a) channel name, b) input parameters (e.g. i_1, i_2, ..., i_m) and c) output parameters (e.g. o_1, o_2, ..., o_n), as shown in Figure 4-3.

$$\boxed{\text{Channel\_Name (In } i_1, i_2, ..., i_m \text{ ; Out } o_1, o_2, ..., o_n)}$$

Figure 4-3   Channel Formula

4-2 Channel-Based Interactions

An interaction represents an indivisible and instantaneous communication or handshake between two agents. In the channel-based approach as shown in Figure 4-4, the caller agent interacts with the callee agent through the channel interaction.

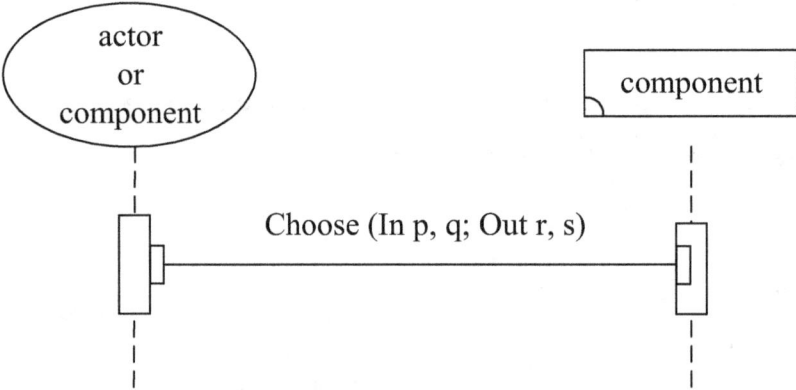

Figure 4-4   Channel-Based Value-Passing Interaction

The caller agent owns the "calling port" of the interaction. In this case, the calling port is " Choose (In p, q; Out r, s) " and its conduct is to assist the caller agent to output a value to each of the "p" and "q" variables (of the "Choose" channel), and input a value from each of the "r" and "s" variables (of the "Choose" channel), as shown in Figure 4-5.

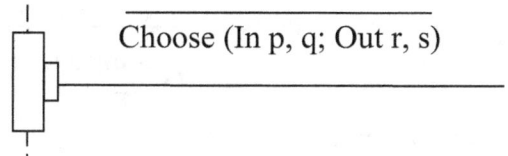

Figure 4-5   Calling Port

The caller agent together with the "calling port" is named the "calling action" as shown in Figure 4-6.

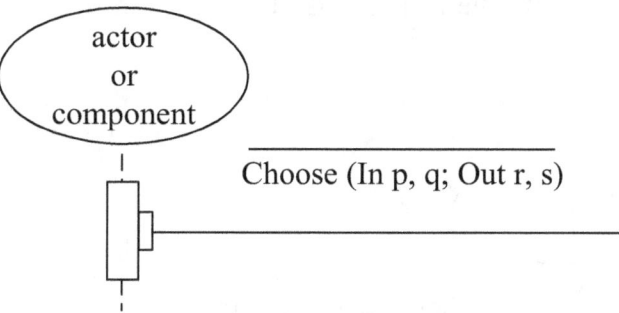

Figure 4-6   Calling Action

The callee agent owns the "called port" of the interaction. In this case, the called port is "Choose (In p, q; Out r, s)" and its conduct is to assist the callee agent to input a value from each of the "p" and "q" variables (of the "Choose" channel), and output a value to each of the "r" and "s" variables (of the "Choose" channel), as shown in Figure 4-7.

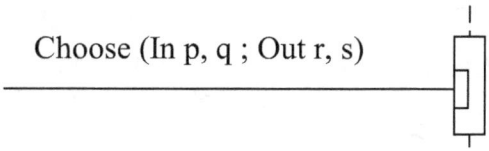

Figure 4-7   Called Port

The callee agent together with the "called port" is named the "called action" as shown in Figure 4-8.

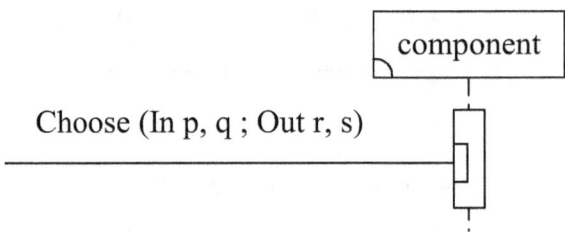

Figure 4-8 Called Action

In order to simplify the channel-based interaction diagram, we will redraw it as shown in Figure 4-9.

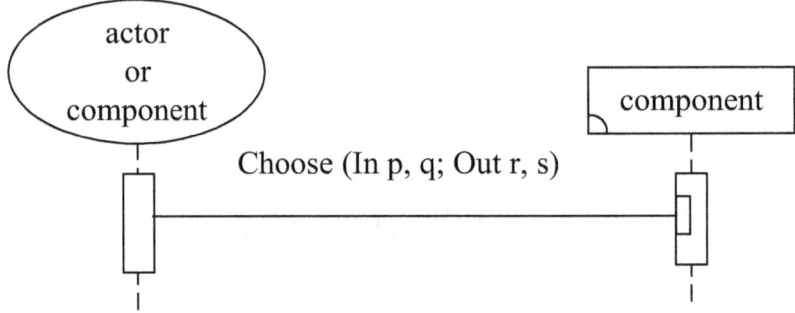

Figure 4-9 Channel-Based Interaction Diagram (I)

Or we can draw the channel-based interaction diagram as shown in Figure 4-10.

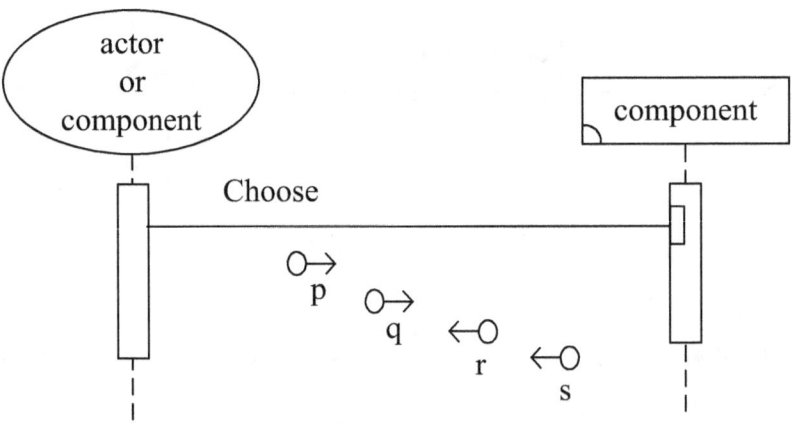

Figure 4-10 Channel-Based Interaction Diagram (II)

We use a channel-based internal interaction (i.e. λ) to represent their handshake or communication, if the caller agent and the callee agent are the same one, as shown in Figure 4-11.

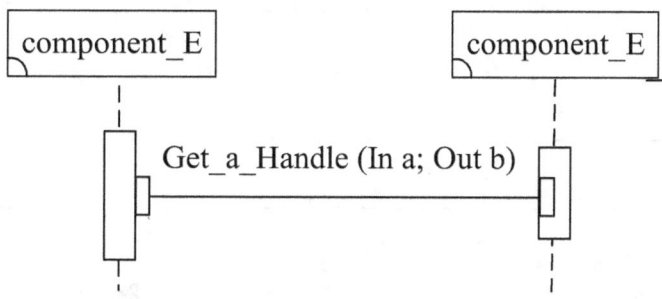

Figure 4-11   An Internal Interaction (I)

Also, we may redraw the internal interaction as shown in Figure 4-12.

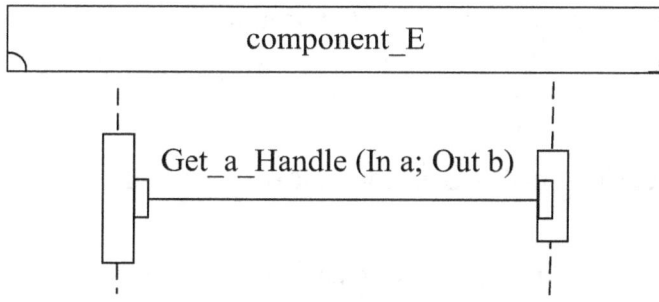

Figure 4-12   An Internal Interaction (II)

4-3 Formal Description of a Channel-Based Communication Port

We formally describe a channel-based communication port as a 2-tuple PORT = <calling_or_called, channel_formula>, where "calling_or_called" stands for a CALLING or CALLED port tag and "channel_formula" stands for a channel formula as shown in Figure 4-13.

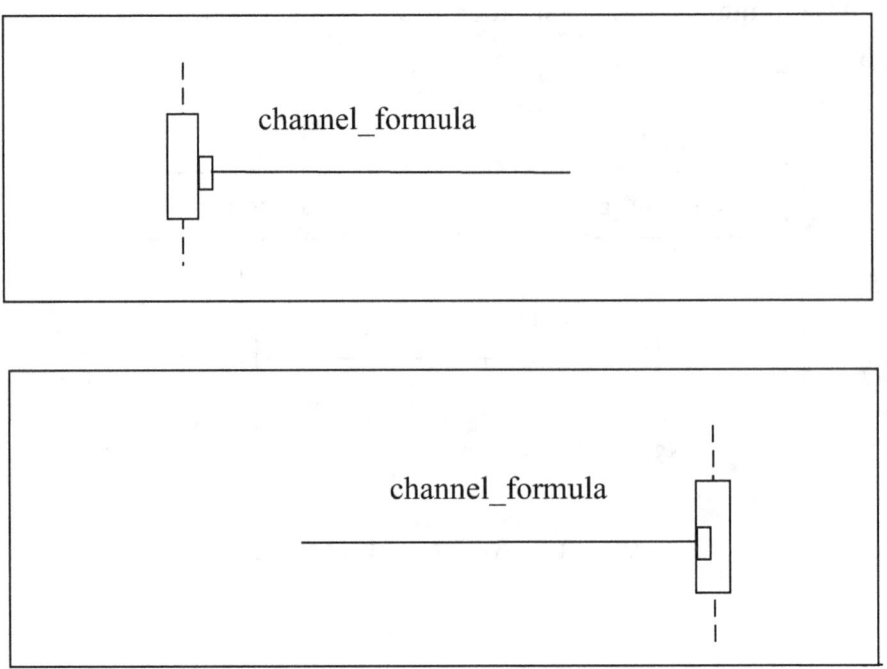

Figure 4-13 Formal Description of
a Channel-Based Communication Port

4-4 Formal Description of a Channel-Based Action

We formally describe a channel-based action as a 3-tuple ACTION = <agent, calling_or_called, channel_formula>, where "agent" stands for the name of a caller or callee agent, "calling_or_called" stands for a CALLING or CALLED action tag, and "channel_formula" stands for a channel formula as shown in Figure 4-14.

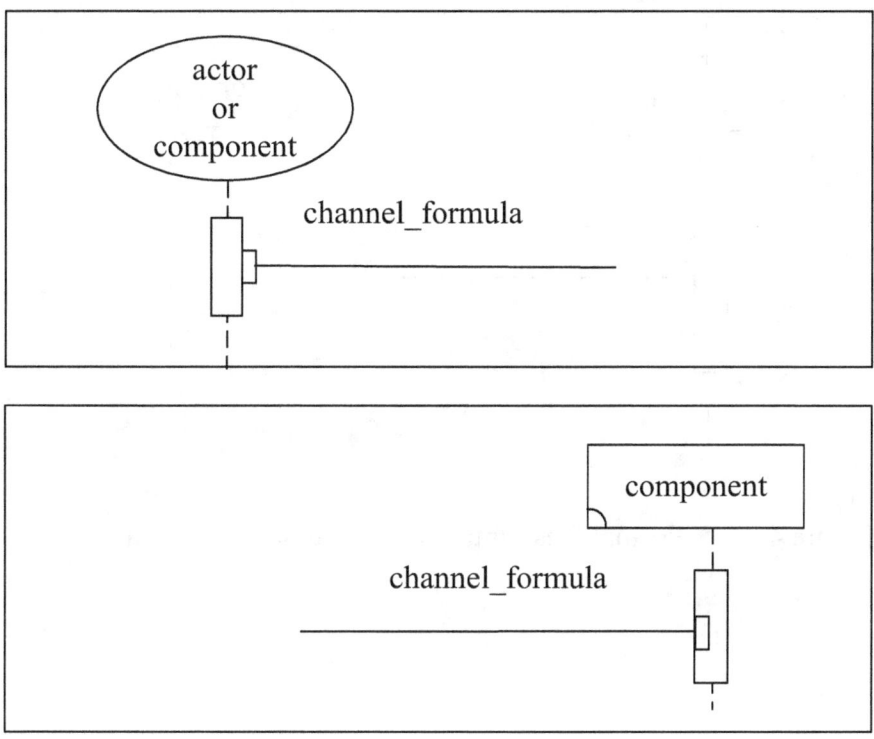

Figure 4-14　Formal Description of a Channel-Based Action

4-5 Formal Description of a Channel-Based Interaction

We formally describe a channel-based interaction as a 3-tuple INTERACTION = <caller_agent, channel_formula, callee_agent>, where "caller_agent" stands for the name of a caller agent, "channel_formula" stands for a channel formula and "callee_agent" stands for the name of a callee agent as shown in Figure 4-15.

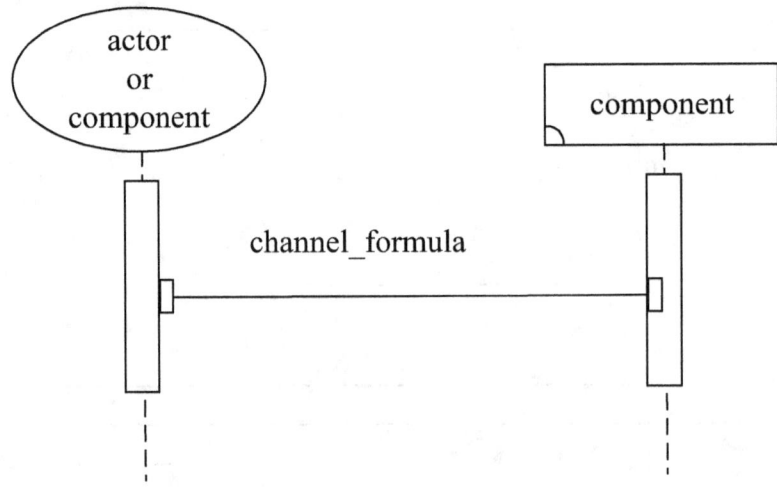

Figure 4-15   Formal Description of a Channel-Based Interaction

Chapter 5: The Structure-Behavior Coalescence Approach

Systems structure and systems behavior are the two most prominent views of a system, integrating the systems structure and systems behavior is apparently the best way to achieve an integrated whole of a system. If we are not able to integrate the systems structure and systems behavior, then there is no way that we are able to integrate the whole system. Structure-behavior coalescence (SBC) provides an elegant way to integrate the systems structure and systems behavior of a system. In other words, SBC facilitates an integrated whole of a system.

5-1 Structure-Behavior Coalescence Means to Integrate the Systems Structure and Systems Behavior

Systems structure, specified by components, their channels and their composition, refers to the type of connection between the components of a system as shown in Figure 5-1.

Figure 5-1 Systems Structure

Systems behavior, specified by the interactions between and among the components and environment, refers to the interconnectivities a system in conjunction with its environment as shown in Figure 5-2.

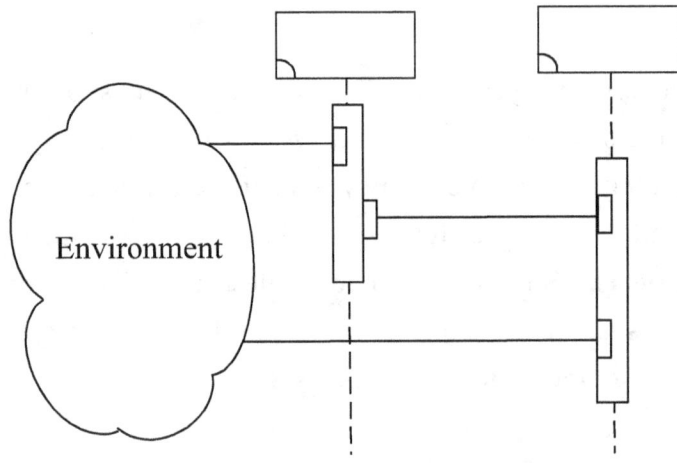

Figure 5-2   Systems Behavior

Systems structure and systems behavior are the two most prominent views of a system, integrating the systems structure and systems behavior is apparently the best way to achieve an integrated whole of a system.

If we are not able to integrate the systems structure and systems behavior, then there is no way that we are able to integrate the whole system.

Structure-behavior coalescence (SBC) [Chao15a] provides an elegant way to integrate the systems structure and systems behavior of a system. In other words, SBC facilitates an integrated whole of a system.

5-2 Interactions among Components and the External Environment to Draw Forth the Systems Behavior

All things that strike us as something independent are essentially parts of a system. We usually call the parts of a system its components. Components are sometimes labeled as parts, entities, objects, building blocks and non-aggregated systems [Chao14a, Chao14b, Chao14c, Chao16a].

In a system, if the components, and among them and the external environment to interact (or handshake), such interaction will draw forth the systems behavior.

A component uses an "action" to interact with the external environment as shown in Figure 5-3.

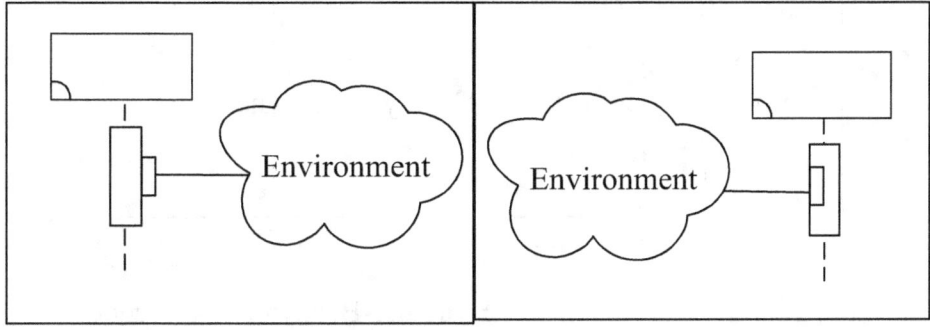

Figure 5-3   A Component Uses an "Action" to Interact
with the External Environment

Two components use an "interaction" to interact with each other as shown in Figure 5-4.

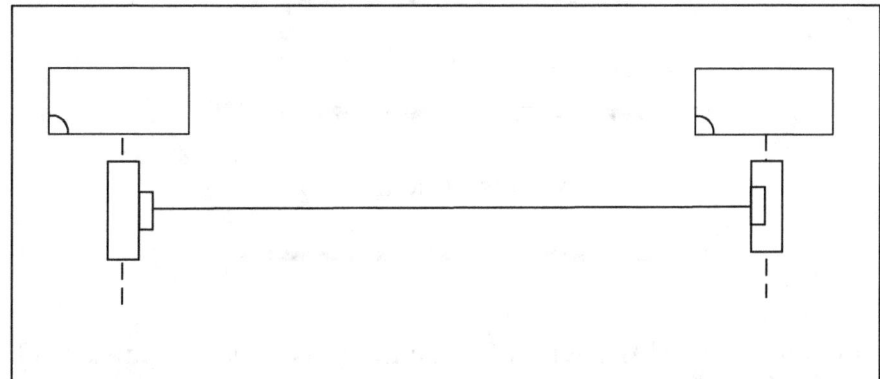

Figure 5-4   Two Components Use an "Interaction"
to Interact with Each Other

5-3 Core Theme of Structure-Behavior Coalescence

The core theme of structure-behavior coalescence is: "Systems Architecture = Systems Structure -->> Systems Behavior." That is, the systems structure will lead to the systems behavior as shown in Figure 5-5.

```
┌─────────────────────────────────────────────────┐
│                                                 │
│     Systems Structure X  ──>>  Systems Behavior X │
│                                                 │
└─────────────────────────────────────────────────┘
```

Figure 5-5   Core Theme of Structure-Behavior Coalescence

One systems structure will draw forth one systems behavior. That is, the systems behavior is attached to or built on the systems structure in the SBC approach.

In other words, the systems behavior can not exist alone; it must be loaded on the systems structure just like a cargo is loaded on a ship as shown in Figure 5-6.

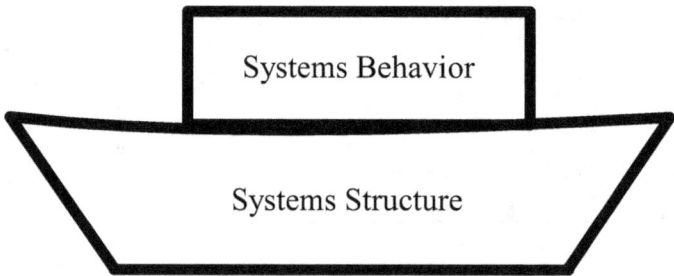

Figure 5-6   Systems Behavior Must be Loaded on the Systems Structure

Chapter 6: Language Constructs of Channel-Based Single-Queue SBC Process Algebra

In the chapter, we illustrate in detail those channel-based single-queue SBC process algebra language constructs which make up the system modeling of a system.

6-1 Entity Set and Entity Name

As shown in Figure 6-1, we assume an infinite set K of channel formulas, and use k_1, k_2...to range over K. Further, we let U be the set of called actions, and use u_1, u_2...to range over U. We let Δ be the set of non-internal interactions, and use a_1, a_2...to range over Δ. We let O be the set of called actions or non-internal interactions, and use o_1, o_2... to range over O. We let T be the set of [condition](called actions/non-internal interactions)[variable settings], and use t_1, t_2...to range over T. Further, we let X be the set of process variables, and use X_1, X_2...to range over X. We let Φ be the set of process Constants, and use A_1, A_2...to range over Φ. We let Π be the set of processes, and use P_1, Q_1...to range over Π. We let Ψ be the set of process expressions, and use E_1, E_2... to range over Ψ. Finally, we let Γ be the set of components, and use C_1, C_2...to range over Γ.

Entity set	Entity name	Type of entity
K	$k_1, k_2...$	channel formulas
U	$u_1, u_2...$	called actions
	λ	internal interaction
Δ	$a_1, a_2...$	non-internal interactions
O	$o_1, o_2...$	called actions or non-internal interactions (called actions/non-internal interactions)
T	$t_1, t_2...$	[condition](called actions/non-internal interactions) [variable settings]
X	$X_1, X_2...$	process variables
Φ	$A_1, A_2...$	process Constants
	I, J,...	indexing sets
Π	$P_1, Q_1...$	processes
Ψ	$E_1, E_2...$	process expressions
Γ	$C_1, C_2...$	components

Figure 6-1   Entities

6-2 Backus-Naur Form of Channel-Based Single-Queue SBC Processes

The set of channel-based single-queue SBC (i.e. Structure-Behavior Coalescence) processes is defined by the following BNF grammar, as shown in Figure 6-2.

(1) $E ::= $ **"fix("** <Process_Variable> **"="**<IFD>{**"+"** <IFD>} **")"**

(2) <IFD> ::= <Type_1_Expression>
    {"●" <Type_1_Or_2_Expression>} " ● " <Process_Variable>

(3) <Type_1_Or_2_Expression> ::= <Type_1_Expression>

    | <Type_2_Expression>

(4) <Type_1_Expression> ::= <Called-Action>[variable settings]

    | $\sum_{j \in I}$ **if** $cond_j$ **then** <Called-Action$_j$>[variable settings$_j$]

(5) <Type_2_Expression> ::= <Interaction>[variable settings]

    | $\sum_{j \in I}$ **if** $cond_j$ **then** <Interaction$_j$>[variable settings$_j$]

Figure 6-2   Backus-Naur Form
of Channel-Based Single-Queue SBC Processes

6-2-1 Recursion of Summation of One or More Interaction Flow Diagrams Defines the Channel-Based Single-Queue SBC Process of a System

Rule 1 describes that the recursion (i.e. **fix**) of summation of one or more interaction flow diagrams (i.e. IFD) defines the channel-based single-queue SBC process of a system, as shown in Figure 6-3.

Rule 1
$E ::= $ **"fix("** <Process_Variable> **"="**<IFD>{**"+"** <IFD>} **")"**

Figure 6-3   Rule 1

6-2-2 An Interaction Flow Diagram Consists of a Type_1 Expression, Followed by Zero or More Type_1_Or_2 Expressions

Rule 2 describes that an interaction flow diagram (i.e. IFD) consists of a type_1 expression (i.e. Type_1_Expression) and followed by zero or more type_1 or type_2 expressions (i.e. Type_1_Or_2_Expression), as shown in Figure 6-4.

Rule 2
<IFD> ::= <Type_1_Expression> {"●" <Type_1_Or_2_Expression>} "●" <Process_Variable>

Figure 6-4   Rule 2

6-2-3 Type_1_Or_2 Expression is either Type_1 or Type_2

Rule 3 describes that the type_1_or_2 expression (i.e. Type_1_Or_2_Expression) is either a type_1 expression (i.e. Type_1_Expression) or a type_2 expression (i.e. Type_2_Expression), as shown in Figure 6-5.

Rule 3
<Type_1_Or_2_Expression> ::= <Type_1_Expression> 　　| <Type_2_Expression>

Figure 6-5   Rule 3

6-2-4 Type_1 Expression is either a Called Action with Optional Variable Settings or a Conditional Expression of Called Action with Optional Variable Settings

Rule 4 describes that the type_1 expression (i.e. Type_1_Expression) is either a called action with optional variable settings (i.e. Called-Action[variable settings]) or a conditional expression of called action with optional variable settings (i.e. one-or-more-armed conditional expression of Called-Action$_j$[variable settings$_j$]), as shown in Figure 6-6.

Rule 4
<Type_1_Expression> ::=   <Called-Action>[variable settings] $\quad\mid\sum_{j\in I}$ **if** $cond_j$ **then** <Called-Action$_j$>[variable settings$_j$]

Figure 6-6   Rule 4

6-2-5 Type_2 Expression is either an Interaction with Optional Variable Settings or a Conditional Expression of Interaction with Optional Variable Settings

Rule 5 describes that the type_2 expression (i.e. Type_2_Expression) is either an interaction with optional variable settings (i.e. Interaction[variable settings]) or a conditional expression of interaction with optional variable settings (i.e. one-or-more-armed conditional expression of Interaction$_j$[variable settings$_j$]), as shown in Figure 6-7.

Rule 5
<Type_2_Expression> ::=   <Interaction>[variable settings] $\quad\mid\sum_{j\in I}$ **if** $cond_j$ **then** <Interaction$_j$>[variable settings$_j$]

Figure 6-7   Rule 5

Chapter 7: Transitional Semantics of Channel-Based Single-Queue SBC Process Algebra

In the chapter, we illustrate in detail those channel-based single-queue SBC process algebra transitional semantics which regards the system modeling of a system.

7-1 Transitional Semantics

In giving meaning to the channel-based single-queue SBC process algebra, we shall use the following labelled transition system (LTS)

$$(\Psi, T, \rightarrow)$$

which consists of a set Ψ of process expressions, a set T of "[condition](called actions/non-internal interactions)[variable settings]", and a transition relation $\rightarrow \subseteq \Psi \times T \times \Psi$ where $(E_i, t, E_j) \in \rightarrow$ is denoted by $E_i \xrightarrow{t} E_j$.

The semantics for Ψ consists in the transition rules of each transition relation $\rightarrow$ over $\Psi \times T \times \Psi$. These transition rules will follow the structure of expressions.

As shown in Figure 7-1, we give the complete set of transition rules; the names Prefix, Sum, Recursion, and Constant indicate that the rules are associated respectively with Prefix, Summation, and Recursion and with Constants.

$$\text{Prefix} \quad \dfrac{}{t \bullet E \xrightarrow{t} E}$$

$$\text{Sum}_j \quad \dfrac{E_j \xrightarrow{t} E'_j}{\sum_{i \in I} E_i \xrightarrow{t} E'_j} (j \in I)$$

$$\text{Recursion} \quad \dfrac{\mathbf{fix}(X=z\{\mathbf{fix}(X=z)/X\}) \xrightarrow{t} E'}{\mathbf{fix}(X=z) \xrightarrow{t} E'}$$

$$\text{Constant} \quad \dfrac{P \xrightarrow{t} P'}{A \xrightarrow{t} P'} (A \stackrel{\text{def}}{=} P)$$

Figure 7-1 Transition Rules for
the Channel-Based Single-Queue SBC Process Algebra

7-2 Rule of Prefix

The rule for Prefix, shown in Figure 7-2, can be read as follows: Under any circumstances, we always infer $t \bullet E \xrightarrow{t} E$. That is, an expression, with a [condition](called action/non-internal interaction)[variable settings] prefixed to it, will use this [condition](called action/non-internal interaction)[variable settings] to accomplish the transition.

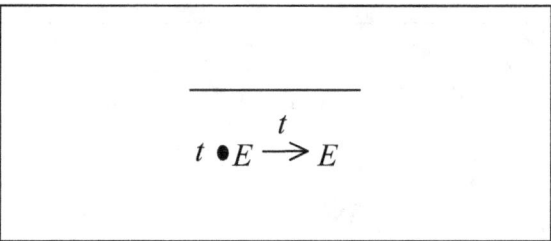

Figure 7-2   Rule of Prefix

7-3 Rule of Summation

The rule for Summation, shown in Figure 7-3, can be read as follows: If any one summand E_j of the sum $\sum_{i \in I} E_i$ has a [condition](called action/non-internal interaction)[variable settings], then the whole sum also has that [condition](called action/non-internal interaction)[variable settings].

$$\frac{E_j \xrightarrow{t} E'_j}{\sum_{i \in I} E_i \xrightarrow{t} E'_j} \; (j \in I)$$

Figure 7-3   Rule of Summation

7-4 Rule of Recursion

The rule for Recursion, shown in Figure 7-4, can be read as follows: This says that any [condition](called action/non-internal interaction)[variable settings] which may be inferred for the **fix** expression 'unwound' once (by substituting itself for its bound variable) may be inferred for the **fix** expression itself.

$$\frac{\mathbf{fix}(X=z\{\mathbf{fix}(X=z)/X\}) \overset{t}{\rightarrow} E'}{\mathbf{fix}(X=z) \overset{t}{\rightarrow} E'}$$

Figure 7-4   Rule of Recursion

7-5 Rule of Constants

The rule for Constants, shown in Figure 7-5, can be read as follows: the rule of Constants asserts that each Constant has the same transitions as its defining expression.

$$\frac{P \overset{t}{\rightarrow} P'}{A \overset{t}{\rightarrow} P'} \quad (A \overset{\text{def}}{=} P)$$

Figure 7-5   Rule of Constants

PART III: CASES STUDY

Chapter 8: Channel-Based Single-Queue SBC Process of the Robot System

In this chapter, we use the channel-based single-queue SBC process algebra to model the *Robot* system as shown in Figure 8-1.

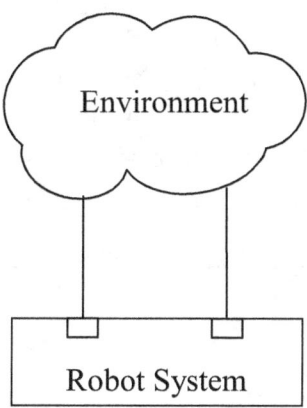

Figure 8-1   Systems Modeling the *Robot* System

8-1 BNF Tree of the Robot System

The channel-based single-queue SBC process of the *Robot* system, A_{001}, is defined as "**fix**$(X_{001}=t_{001}\bullet t_{002}\bullet X_{001}+t_{003}\bullet t_{004}\bullet X_{001})$".

We draw the channel-based single-queue SBC process algebra Backus-Naur Form tree of A_{001} as shown in Figure 8-2.

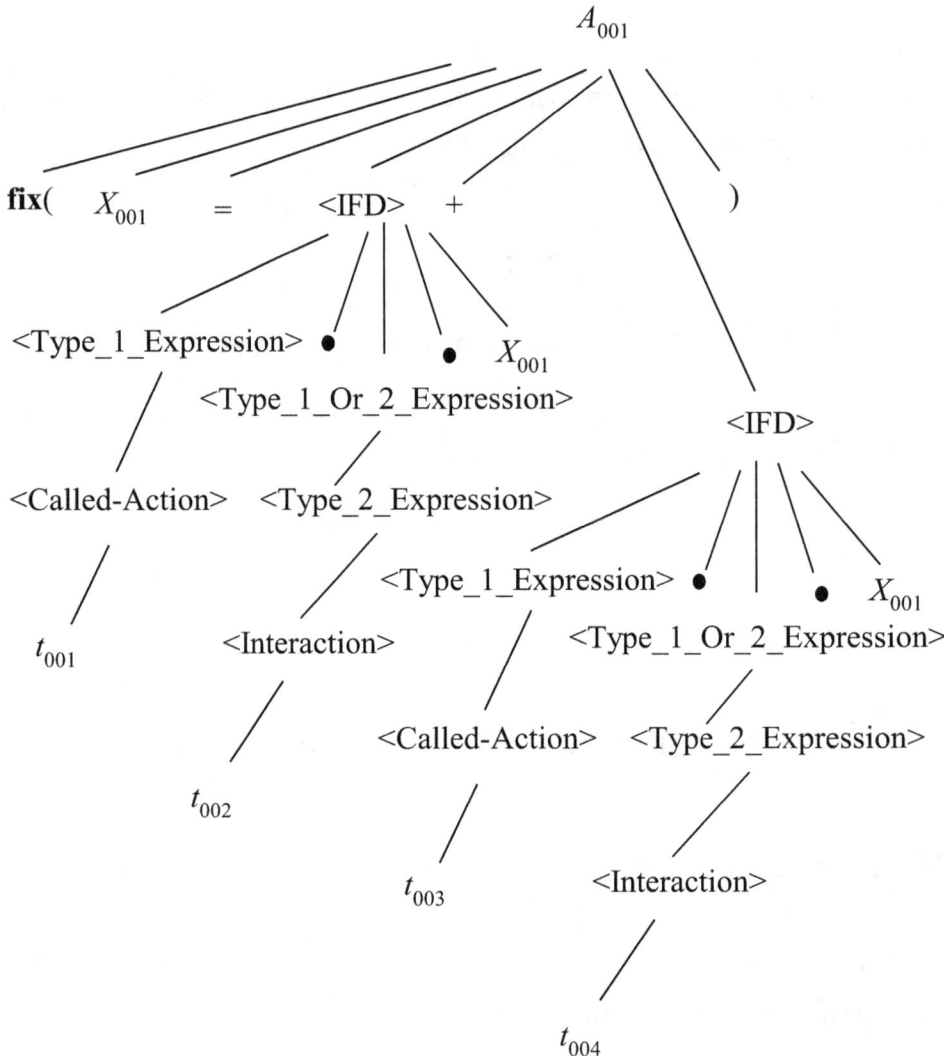

Figure 8-2 Backus-Naur Form Tree of the *Robot* System's
Channel-Based Single-Queue SBC Process

There are two IFDs in the channel-based single-queue SBC process of the *Robot* system. The first IFD is shown in Figure 8-3.

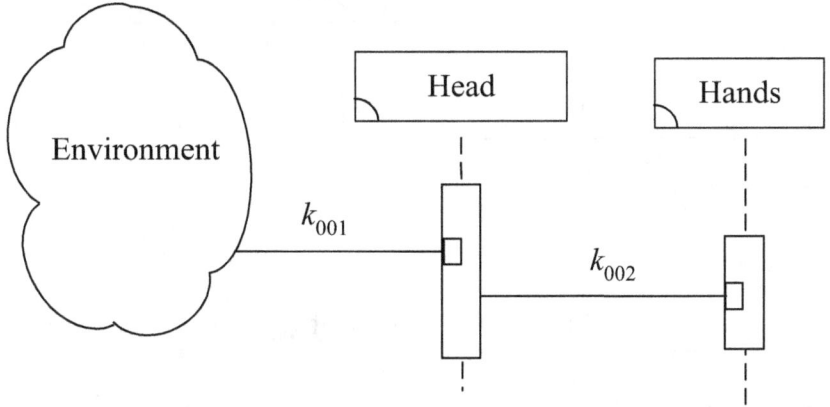

Figure 8-3　First IFD of the *Robot* System

The second IFD of the channel-based single-queue SBC process of the *Robot* system is shown in Figure 8-4.

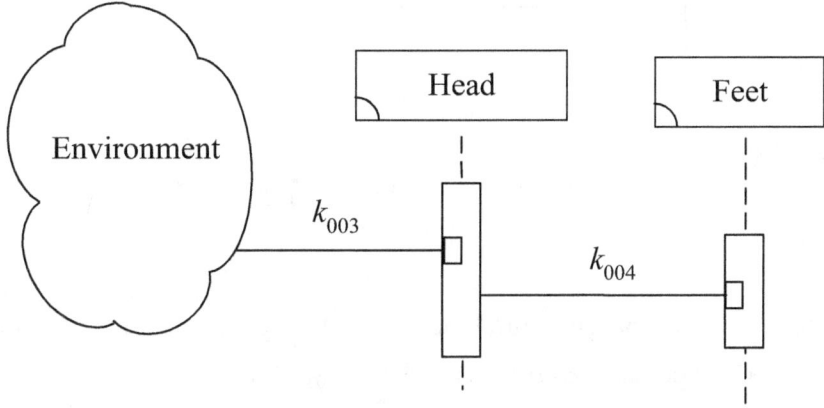

Figure 8-4　Second IFD of the *Robot* System

8-2 Prefixes of the Robot System

The t_{001} (channel-based called action with no conditions and no variable settings) prefix is defined as "<*Head*, CALLED, k_{001}>", as shown in Figure 8-5.

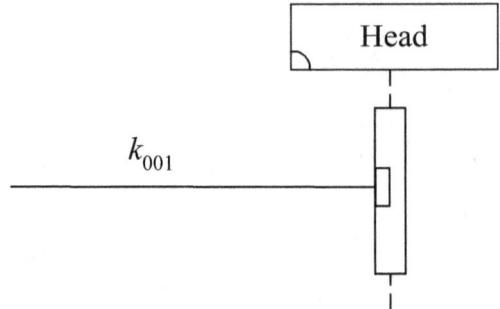

Figure 8-5   Prefix of t_{001}

The t_{002} (channel-based interaction with no conditions and no variable settings) prefix is defined as "<*Head*, k_{002}, *Hands*>", as shown in Figure 8-6.

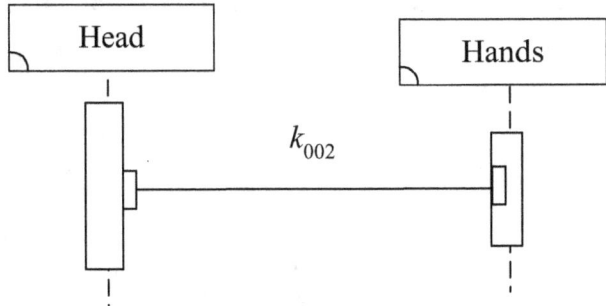

Figure 8-6   Prefix of t_{002}

The t_{003} (channel-based called action with no conditions and no variable settings) prefix is defined as "<*Head*, CALLED, k_{003}>", as shown in Figure 8-7.

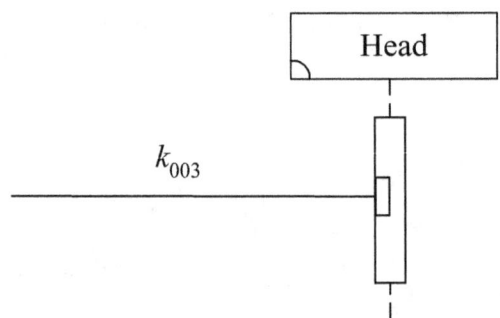

Figure 8-7   Prefix of t_{003}

The t_{004} (channel-based interaction with no conditions and no variable settings) prefix is defined as "<*Head*, k_{004}, *Feet*>", as shown in Figure 8-8.

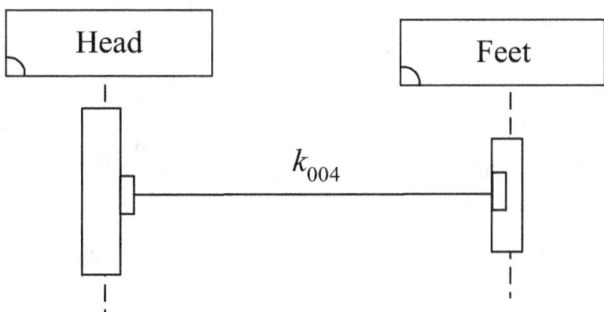

Figure 8-8   Prefix of t_{004}

Figure 8-9 shows all channel formulas of the channel-based single-queue SBC process of the *Robot* system.

Entity name	Channel Formula
k_{001}	Receive_Write_Signal
k_{002}	Move_Hand
k_{003}	Receive_Walk_Signal
k_{004}	Move_Foot

Figure 8-9   Channel Formulas of the *Robot* System's Process

8-3 Process of the Robot System

The following transition graph shows, in Figure 8-10, the semantics of A_{001}'s channel-based single-queue SBC process.

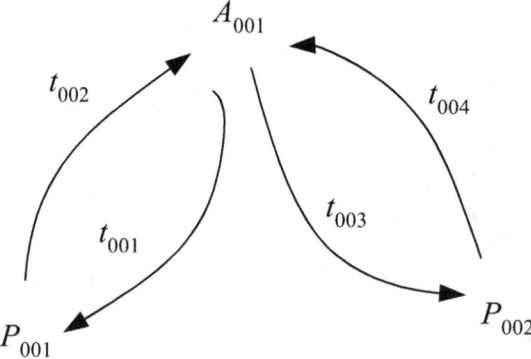

Figure 8-10 Transition graph of the *Robot* System's Process

In the transition graph of the A_{001}'s channel-based single-queue SBC process, processes A_{001}, P_{001} and P_{002} are defined as in Figure 8-11.

$$A_{001} \stackrel{\text{def}}{=} t_{001} \bullet P_{001} + t_{003} \bullet P_{002}$$

$$P_{001} \stackrel{\text{def}}{=} t_{002} \bullet A_{001}$$

$$P_{002} \stackrel{\text{def}}{=} t_{004} \bullet A_{001}$$

Figure 8-11 Definition of Processes A_{001}, P_{001}, and P_{002}

Chapter 9: Channel-Based Single-Queue SBC Process of the Multi-Tier Personal Data System

After the systems development is finished, the *Multi-Tier Personal Data System* shall appear on a multi-tier platform [Wall04] as shown in Figure 9-1.

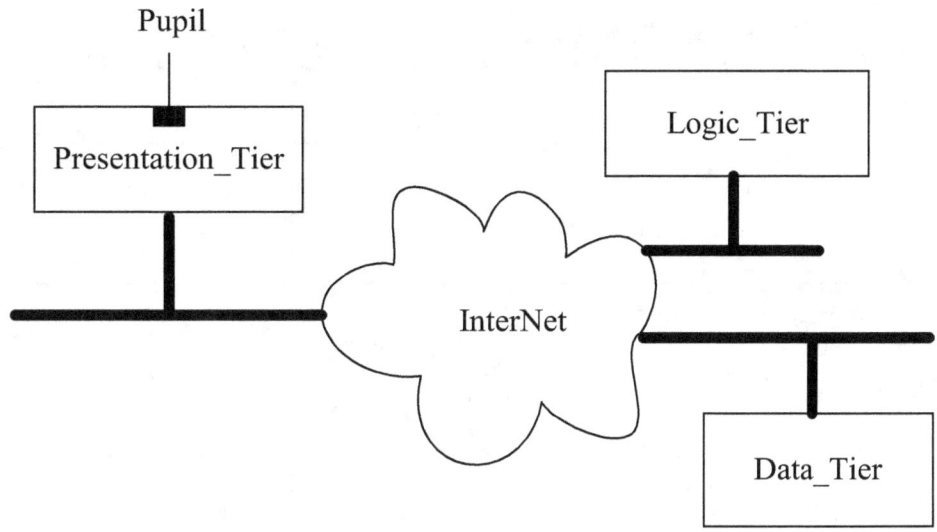

Figure 9-1   *Multi-Tier Personal Data System* on a Multi-Tier Platform

In the *Data_Tier*, there is a *Personal_Database* database [Date03, Elma10] which contains a *Personal_Data* table as shown in Figure 9-2.

Social_Secuirty_Number	Name	Date_of_Birth	Sex	Height (cm)	Weight (Kg)
318-49-2465	Mary R. Williams	June 17, 1976	Female	165	51
424-87-3651	Lee H. Wulf	July 24, 1982	Female	162	76
512-24-3722	John K. Bryant	May 12, 1954	Male	180	80

Figure 9-2   *Personal_Database* Contains *Personal_Data*

The functionality of the *Multi-Tier Personal Data System* is to provide a graphical user interface (GUI) [Gali07] for the external actor to trigger two behaviors. The first behavior is *AgeCalculation* and the second behavior is *OverweightCalculation*, as shown in Figure 9-3.

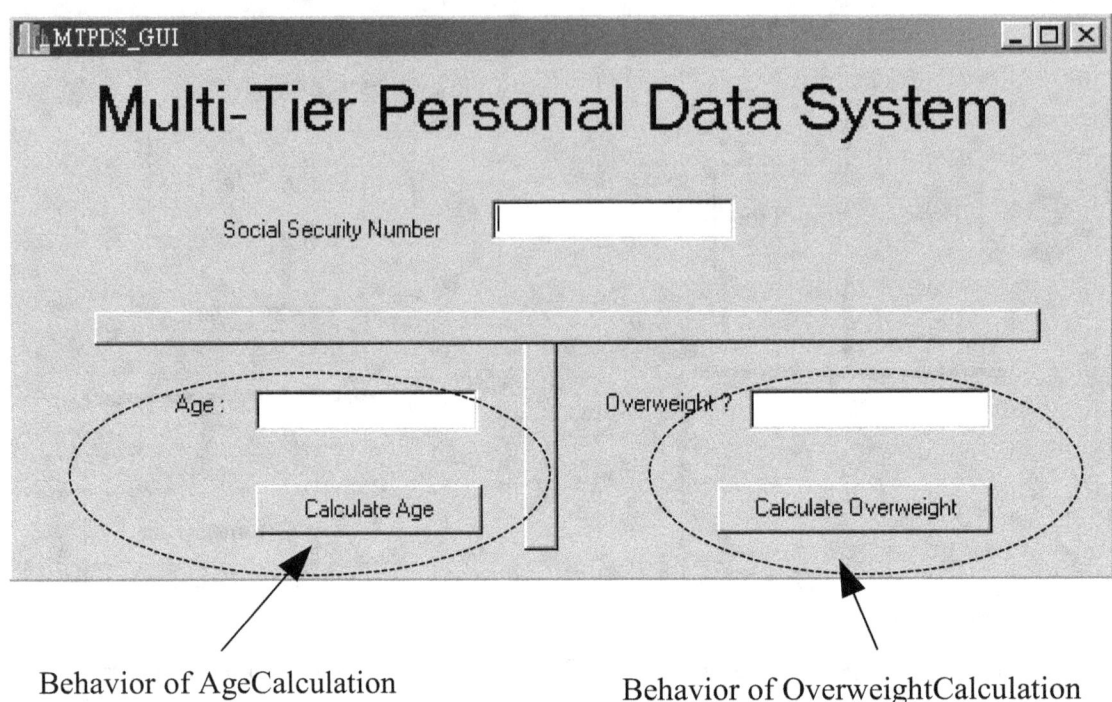

Figure 9-3   Two Behaviors

In the *AgeCalculation* behavior, the external actor inputs an integer *Social_Security_Number* value then presses down the *Calculate_Age* button. After that, the *Multi-Tier Personal Data System* retrieves the *Date_of_Birth* value from the database in line with the corresponding *Social_Security_Number* value. From the *Date_of_Birth* value, the *Multi-Tier Personal Data System* calculates the *Age* value and displays it on the screen. Figure 9-4 shows the *Social_Security_Number* value is 512-24-3722 and the retrieved *Date_of_Birth* value is May 12, 1954 and the calculated *Age* value, which is 59, is then displayed on the screen.

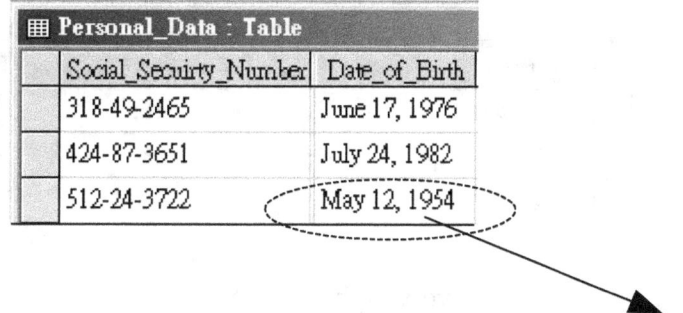

the calculated *Age* value when *Date_of_Birth* is May 12, 1954

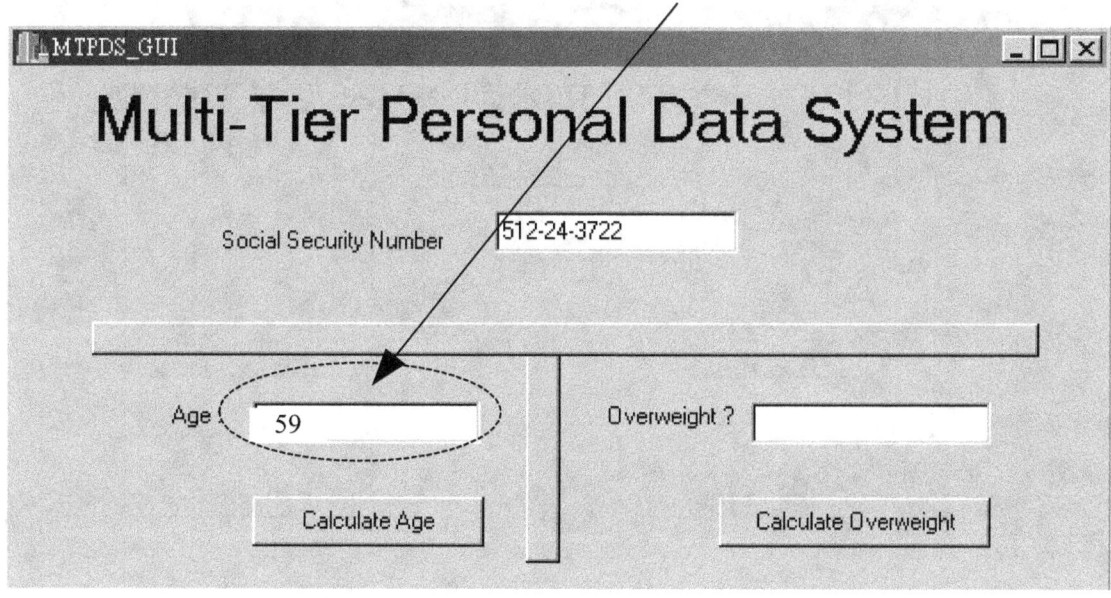

Figure 9-4   Behavior of *AgeCalculation*

In the *OverweightCalculation* behavior, the external actor inputs an integer *Social_Security_Number* value then presses down the *Calculate_Overweight* button. After that, the *Multi-Tier Personal Data System* retrieves the *Weight*, *Height* and *Sex* values from the database in line with the corresponding *Social_Security_Number* value. From the *Weight*, *Height* and *Sex* values, the *Multi-Tier Personal Data System* calculates the true-or-false *Overweight* value and displays it on the screen. Figure 9-5 shows the *Social_Security_Number* value is 318-49-2465 and the retrieved Sex, H*eight* and *Weight* values are Female, 165 and 51, respectively, the calculated *Overweight* value, which is "*No*", is then displayed on the screen.

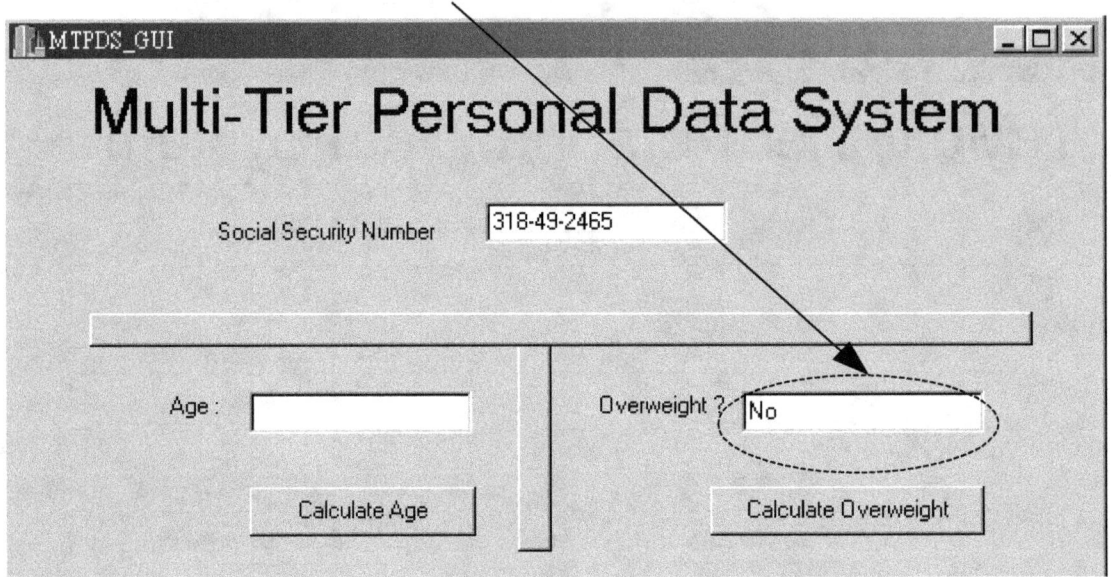

Figure 9-5   Behavior of *OverweightCalculation*

In this chapter, we use the channel-based single-queue SBC process algebra to model the *Multi-Tier Personal Data System* as shown in Figure 9-6.

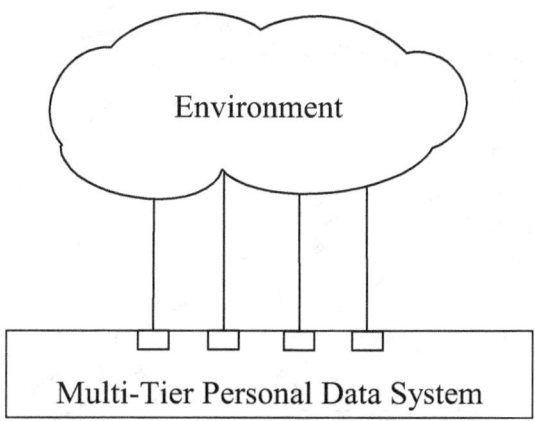

Figure 9-6 Systems Modeling of
the *Multi-Tier Personal Data System*

9-1 BNF Tree of the Multi-Tier Personal Data System

The channel-based single-queue SBC process of the *Multi-Tier Personal Data System*, A_{011}, is defined as "$\mathbf{fix}(X_{011}=t_{011}\bullet t_{012}\bullet t_{013}\bullet t_{014}\bullet t_{015}\bullet X_{011}+t_{016}\bullet t_{017}\bullet t_{018}\bullet t_{019}\bullet t_{020}\bullet X_{011})$".

We draw the channel-based single-queue SBC process algebra Backus-Naur Form tree of A_{011} as shown in Figure 9-7.

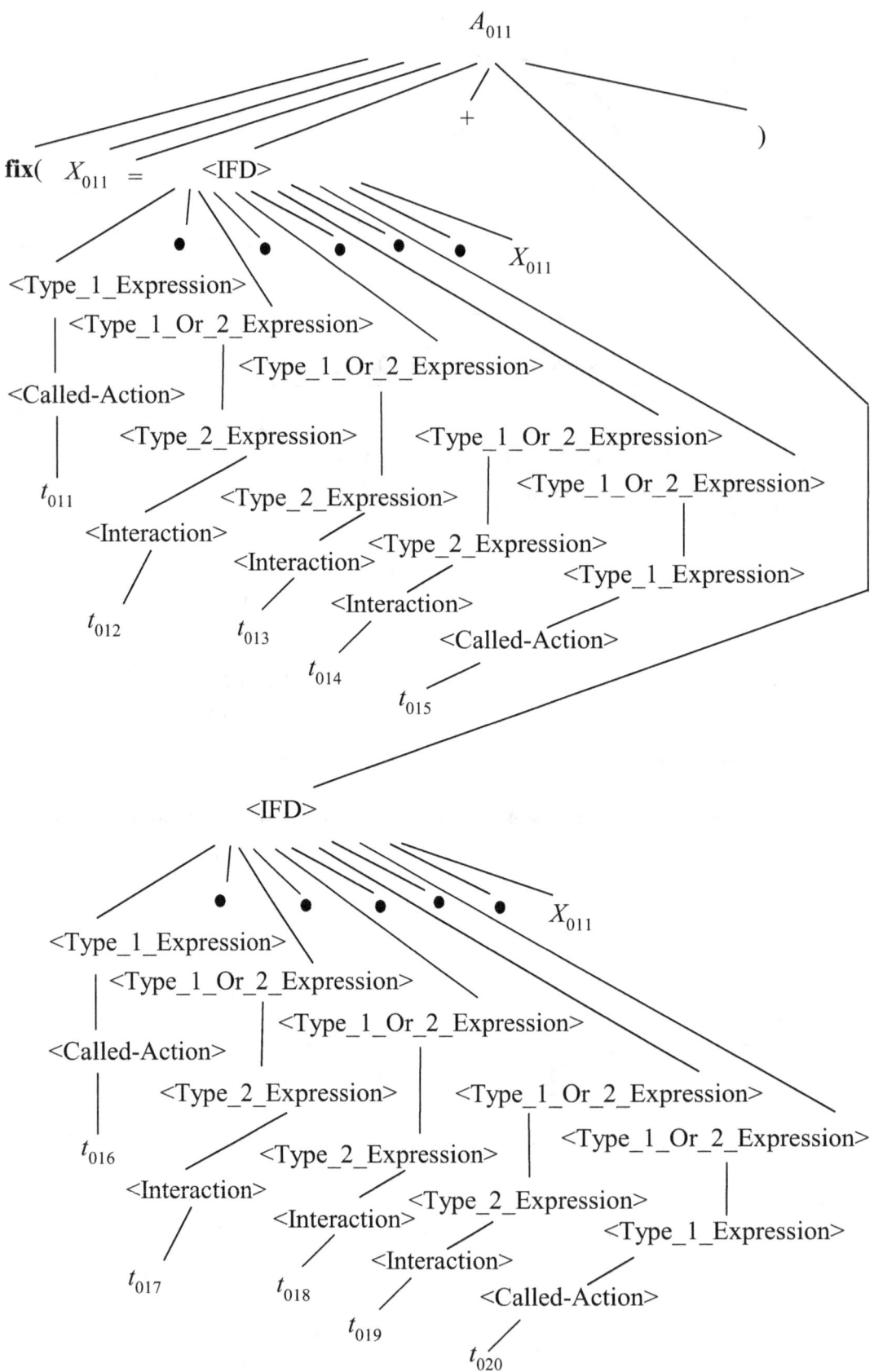

Figure 9-7 Backus-Naur Form Tree of the *Multi-Tier Personal Data System*'s Channel-Based Single-Queue SBC Process

There are two IFDs in the channel-based single-queue SBC process of the *Multi-Tier Personal Data System*. The first IFD is shown in Figure 9-8.

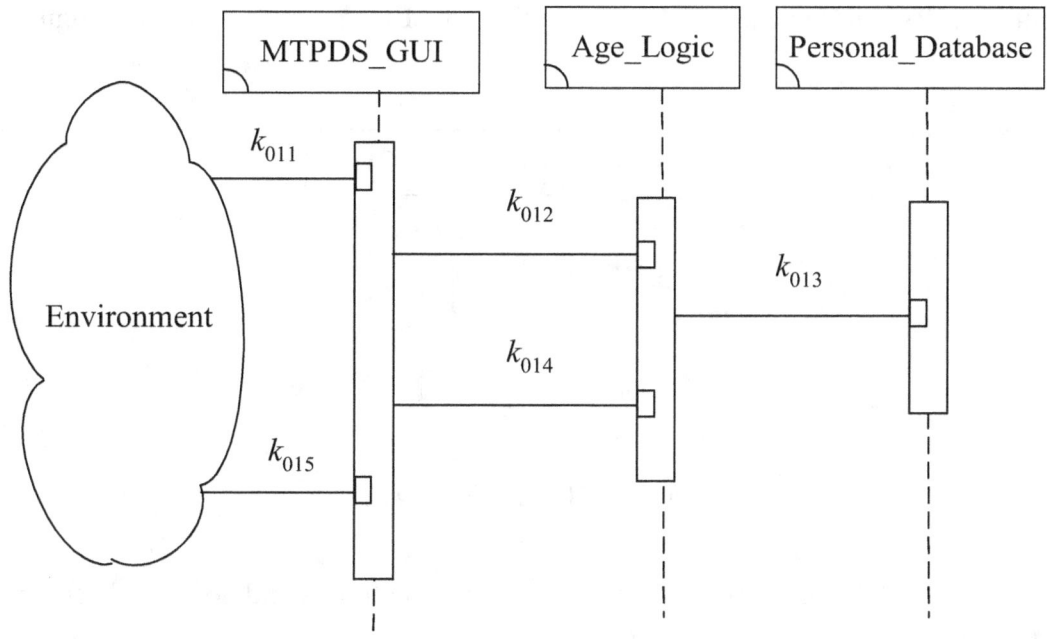

Figure 9-8   First IFD of the *Multi-Tier Personal Data System*

The second IFD of the channel-based single-queue SBC process of the *Multi-Tier Personal Data System* is shown in Figure 9-9.

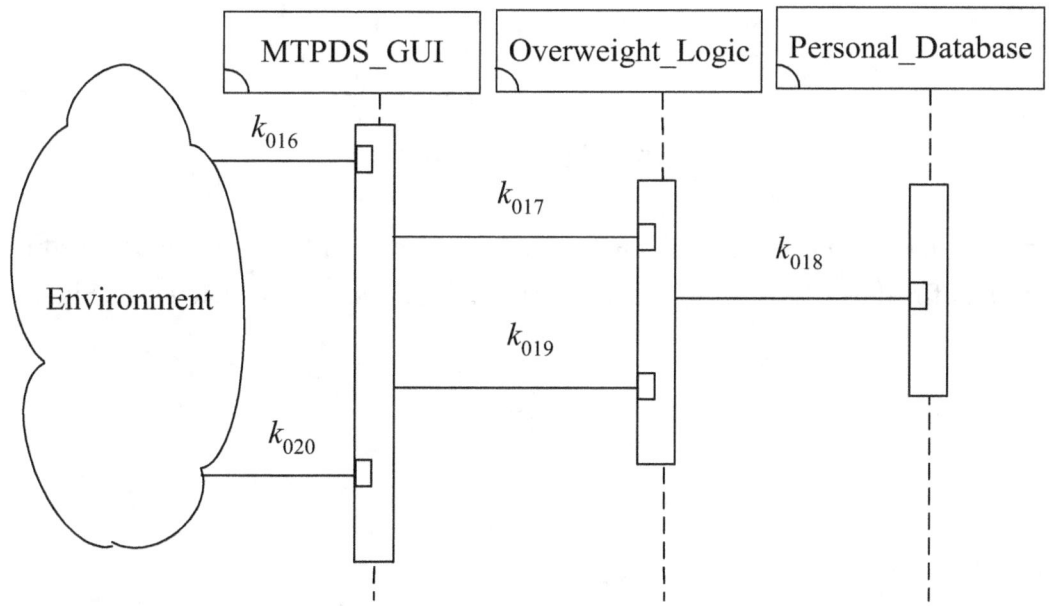

Figure 9-9   Second IFD of the *Multi-Tier Personal Data System*

9-2 Prefixes of the Multi-Tier Personal Data System

The t_{011} (channel-based called action with no conditions and no variable settings) prefix is defined as "<*MTPDS_GUI*, CALLED, k_{011}>", as shown in Figure 9-10.

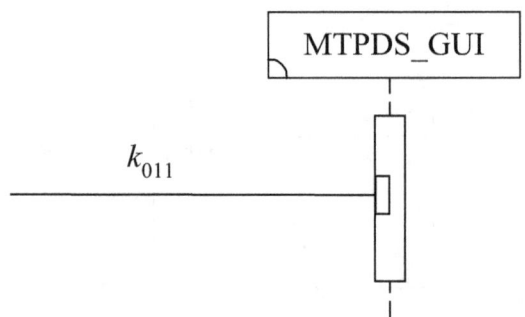

Figure 9-10   Prefix of t_{011}

The t_{012} (channel-based interaction with no conditions and no variable settings) prefix is defined as "<*MTPDS_GUI*, k_{012}, *Age_Logic*>", as shown in Figure 9-11.

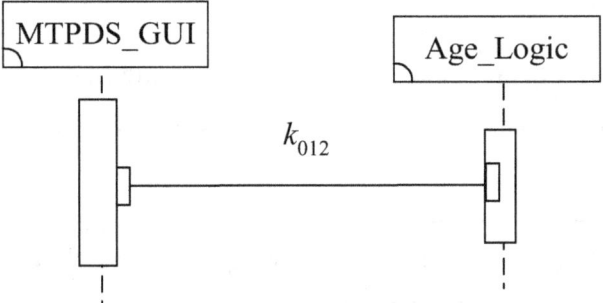

Figure 9-11   Prefix of t_{012}

The t_{013} (channel-based interaction with no conditions and no variable settings) prefix is defined as "<*Age_Logic*, k_{013}, *Personal_Database*>", as shown in Figure 9-12.

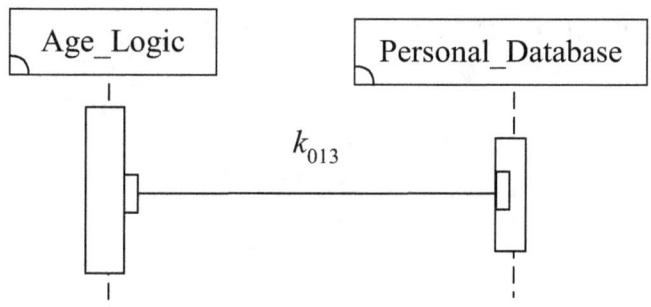

Figure 9-12   Prefix of t_{013}

The t_{014} (channel-based interaction with no conditions and no variable settings) prefix is defined as "<*MTPDS_GUI*, k_{014}, *Age_Logic*>", as shown in Figure 9-13.

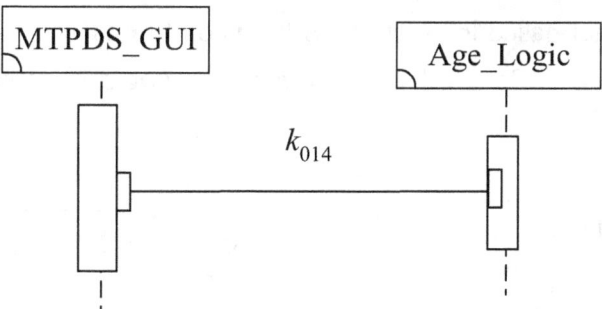

Figure 9-13   Prefix of t_{014}

The t_{015} (channel-based called action with no conditions and no variable settings) prefix is defined as "<*MTPDS_GUI*, CALLED, k_{015}>", as shown in Figure 9-14.

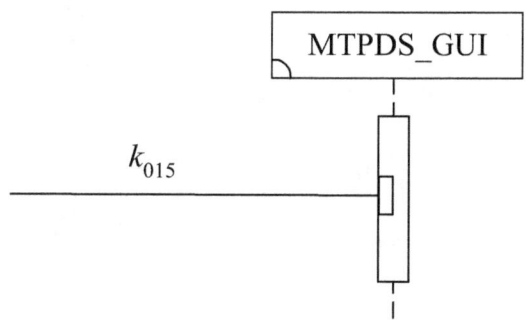

Figure 9-14   Prefix of t_{015}

The t_{016} (channel-based called action with no conditions and no variable settings) prefix is defined as "<*MTPDS_GUI*, CALLED, k_{016}>", as shown in Figure 9-15.

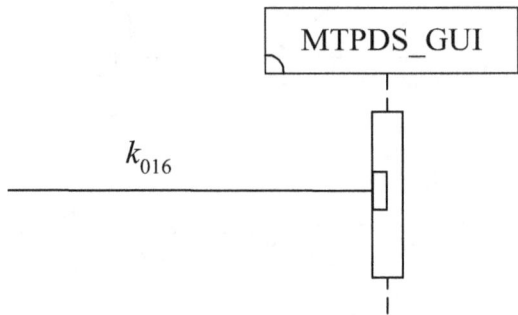

Figure 9-15   Prefix of t_{016}

The t_{017} (channel-based interaction with no conditions and no variable settings) prefix is defined as "<*MTPDS_GUI*, k_{017}, *Overweight_Logic*>", as shown in Figure 9-16.

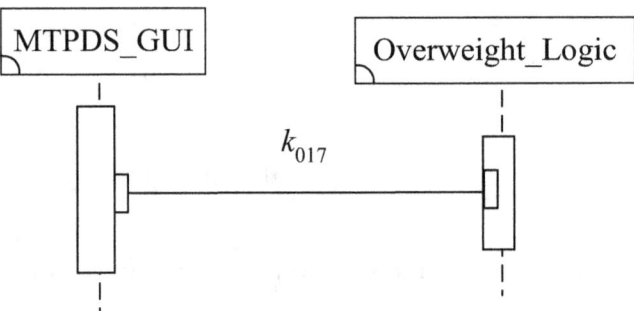

Figure 9-16   Prefix of t_{017}

The t_{018} (channel-based interaction with no conditions and no variable settings) prefix is defined as "<*Overweight_Logic*, k_{018}, *Personal_Database*>", as shown in Figure 9-17.

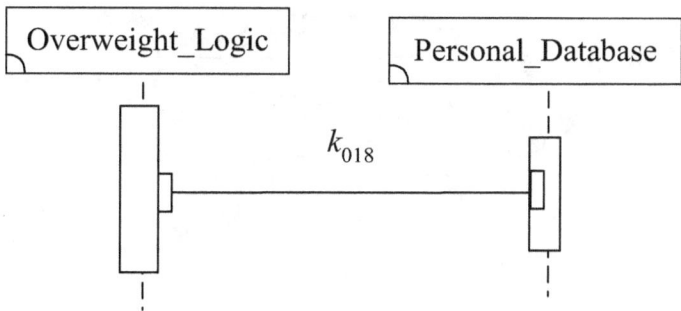

Figure 9-17   Prefix of t_{018}

The t_{019} (channel-based interaction with no conditions and no variable settings) prefix is defined as "<*MTPDS_GUI*, k_{019}, *Overweight_Logic*>", as shown in Figure 9-18.

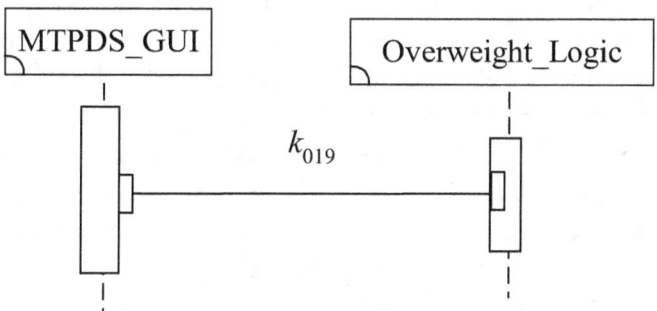

Figure 9-18   Prefix of t_{019}

The t_{020} (channel-based called action with no conditions and no variable settings) prefix is defined as "<*MTPDS_GUI*, CALLED, k_{020}>", as shown in Figure 9-19.

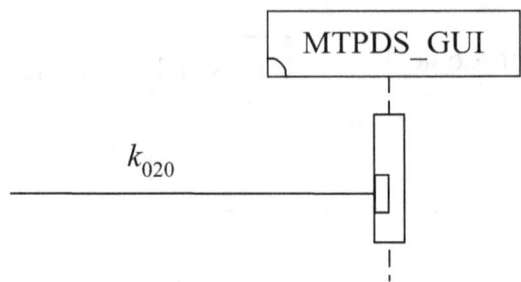

Figure 9-19   Prefix of t_{020}

Figure 9-20 shows all channel formulas of the channel-based single-queue SBC process of the *Multi-Tier Personal Data System*.

Entity name	Channel Formula
k_{011}	Calculate_AgeClick_Call(In Social_Security_Number)
k_{012}	Calculate_Age_Call(In Social_Security_Number)
k_{013}	Sql_DateOfBirth_Select(In Social_Security_Number; Out query_DateOfBirth)
k_{014}	Calculate_Age_Return(Out Age)
k_{015}	Calculate_AgeClick_Return(Out Age)
k_{016}	Calculate_OverweightClick_Call(In Social_Security_Number)
k_{017}	Calculate_Overweight_Call(In Social_Security_Number)
k_{018}	Sql_SexHeightWeight_Select(In Social_Security_Number; Out query_SexHeightWeight)
k_{019}	Calculate_Overweight_Return(Out Overweight)
k_{020}	Calculate_OverweightClick_Return(Out Overweight)

Figure 9-20   Channel Formulas of the *Multi-Tier Personal Data System*'s Process

Figure 9-21 shows the primitive data type specification of the *Social_Security_Number* input parameter and the *Age*, *Overweight* output parameters.

Parameter	Data Type	Instances
Social_Security_Number	Text	424-87-3651, 512-24-3722
Age	Integer	28, 56
Overweight	Boolean	Yes, No

Figure 9-21   Primitive Data Type Specification

Figure 9-22 shows the composite data type specification of the *query_DateOfBirth* output parameter occurring in the *Sql_DateOfBirth_Select(In Social_Security_Number; Out query_DateOfBirth)* channel formula.

Parameter	*query_DateOfBirth*

Data Type	TABLE of
	    Social_Security_Number : Text
	    Age : Integer
	End TABLE;

Instances	424-87-3651 28 512-24-3722 56

Figure 9-22   Composite Data Type Specification

Figure 9-23 shows the composite data type specification of the *query_SexHeightWeight* output parameter occurring in the *Sql_SexHeightWeight_Select(In   Social_Security_Number;   Out query_SexHeightWeight)* channel formula.

Parameter	*query_SexHeightWeight*

Data Type	TABLE of
	   Social_Security_Number : Text
	   Sex : Text
	   Height : Number
	   Weight : Number
	End TABLE;

Instances	424-87-3651 | Female | 162 | 76 512-24-3722 | Male | 180 | 80

Figure 9-23   Composite Data Type Specification

9-3 Process of the Multi-Tier Personal Data System

The following transition graph shows, in Figure 9-24, the semantics of A_{011}'s channel-based single-queue SBC process.

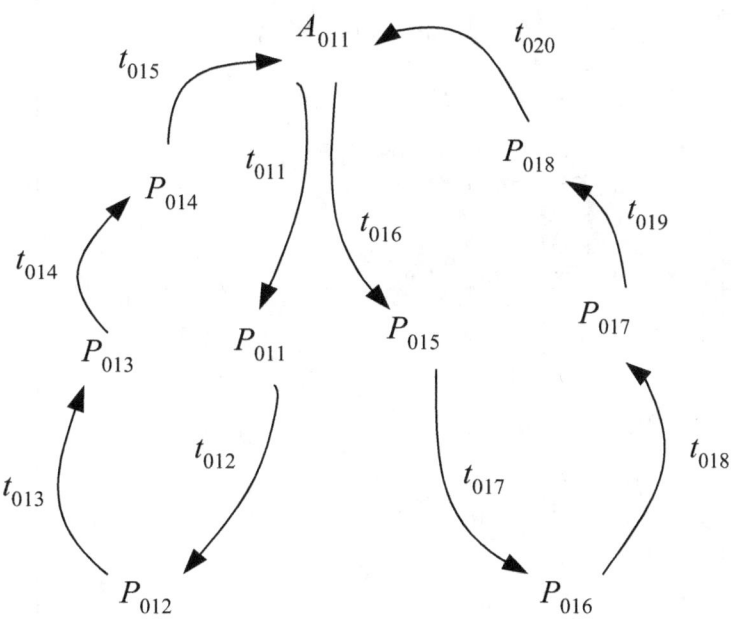

Figure 9-24   Transition graph
of the *Multi-Tier Personal Data System*'s Process

In the transition graph of the A_{011}'s channel-based single-queue SBC process, processes A_{011}, P_{011}, P_{012}, P_{013}, P_{014}, P_{015}, P_{016}, P_{017} and P_{018} are defined as in Figure 9-25.

$$A_{011} \stackrel{def}{=} t_{011} \bullet P_{011} + t_{016} \bullet P_{015}$$

$$P_{011} \stackrel{def}{=} t_{012} \bullet P_{012}$$

$$P_{012} \stackrel{def}{=} t_{013} \bullet P_{013}$$

$$P_{013} \stackrel{def}{=} t_{014} \bullet P_{014}$$

$$P_{014} \stackrel{def}{=} t_{015} \bullet A_{011}$$

$$P_{015} \stackrel{def}{=} t_{017} \bullet P_{016}$$

$$P_{016} \stackrel{def}{=} t_{018} \bullet P_{017}$$

$$P_{017} \stackrel{def}{=} t_{019} \bullet P_{018}$$

$$P_{018} \stackrel{def}{=} t_{020} \bullet A_{011}$$

Figure 9-25 Definition of Processes A_{011}, P_{011}, P_{012}, P_{013}, P_{014}, P_{015}, P_{016}, P_{017}, and P_{018}

Chapter 10: Channel-Based Single-Queue SBC Process of the Mathematical Calculation System

The functionality of the *Mathematical Calculation System* is to provide a graphical user interface (GUI) [Gali07] for the external actor to trigger two behaviors. The first behavior is *DIVIDE&MAXIMUM* and the second behavior is *DIVIDE&GCD*, as shown in Figure 10-1.

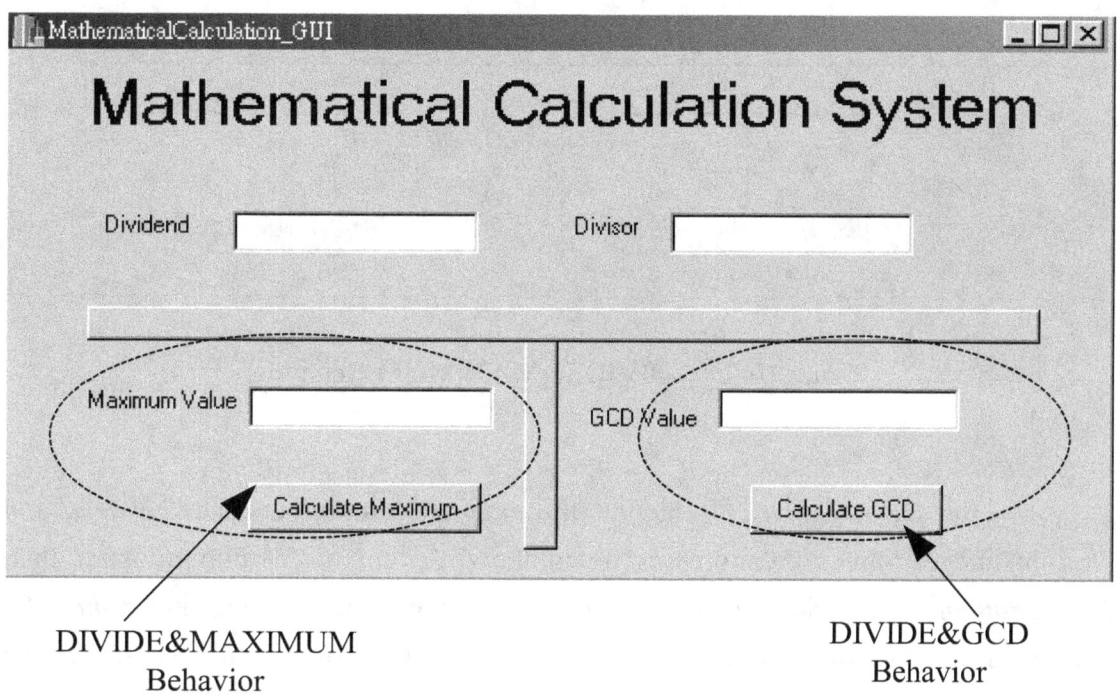

Figure 10-1   Two Behaviors

In the *DIVIDE&MAXIMUM* behavior, the external actor inputs the *Dividend* and *Divisor* integer values then presses down the *Calculate_Maximum* button. After that, *Mathematical Calculation System* will calculate to get the *quotient* and *remainder* values. And from the *quotient* and *remainder* values, *Mathematical Calculation System* selects the maximum one and displays it on the screen. Figure 10-2 shows the *Dividend* value is 88 and the *Divisor* value is 5 and the calculated results of *quotient* and *remainder* respectively are 17 and 3. The maximum value of 17 and 3, which is 17, is then displayed on the screen.

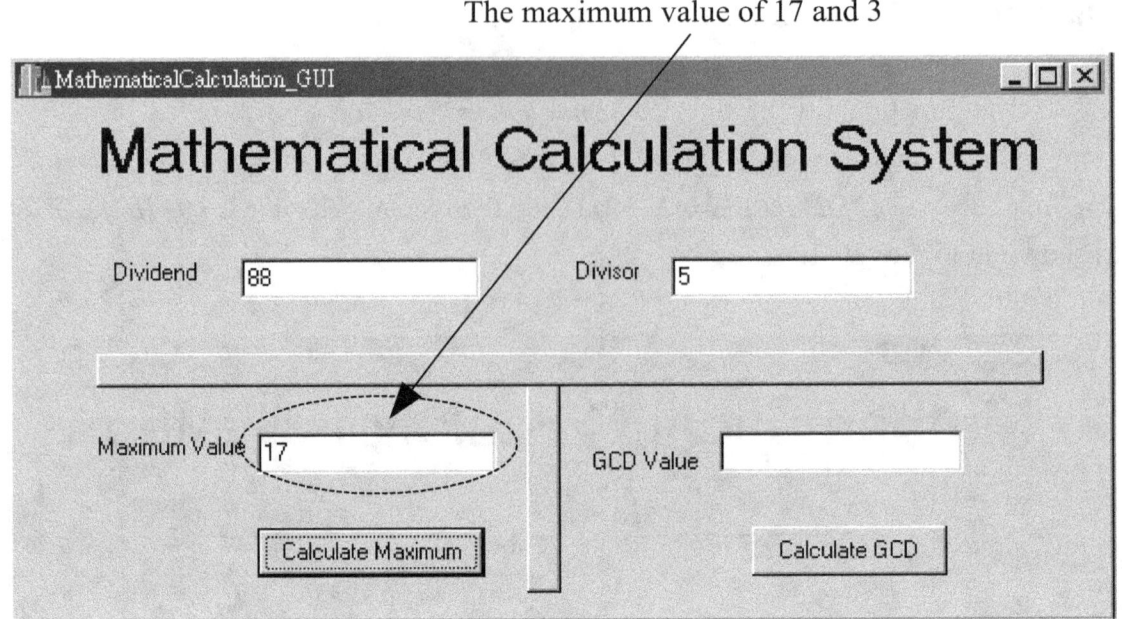

Figure 10-2   DIVIDE&MAXIMUM Behavior

In the *DIVIDE&GCD* behavior, the external actor inputs the *Dividend* and *Divisor* integer values then presses down the *Calculate_GCD* button. After that, *Mathematical Calculation System* will calculate to get the *quotient* and *remainder* values. And from the *quotient* and *remainder* values, *Mathematical Calculation System* will calculate the greatest common divisor value and displays it on the screen. Figure 10-3 shows the *Dividend* value is 224 and the *Divisor* value is 18 and the calculated results of *quotient* and *remainder* respectively are 12 and 8. The greatest common divisor value of 12 and 8, which is 4, is then displayed on the screen.

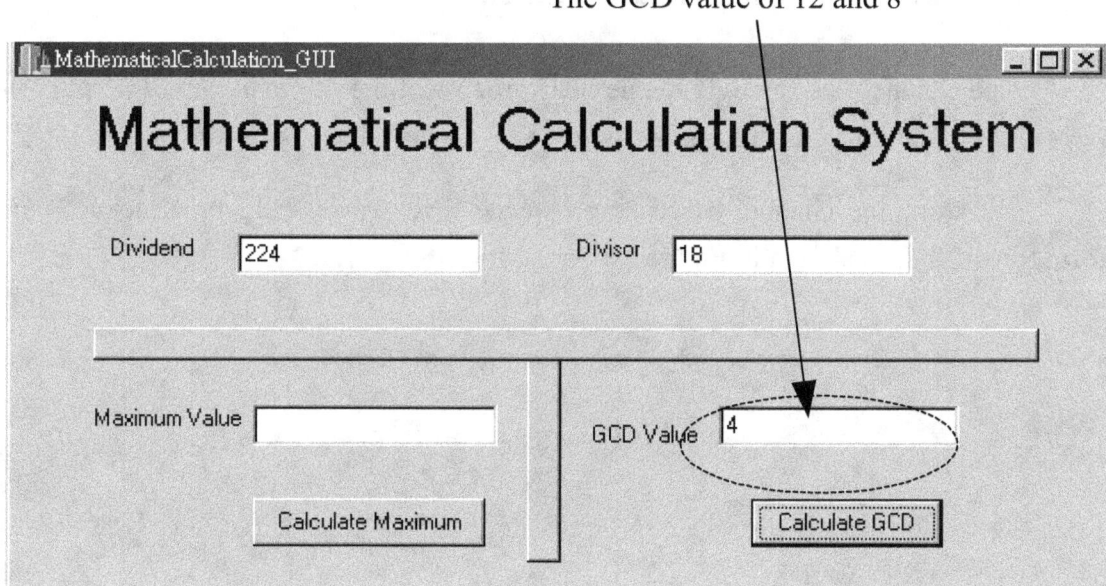

Figure 10-3　　DIVIDE&GCD Behavior

In this chapter, we use the channel-based single-queue SBC process algebra to model the *Mathematical Calculation System* as shown in Figure 10-4.

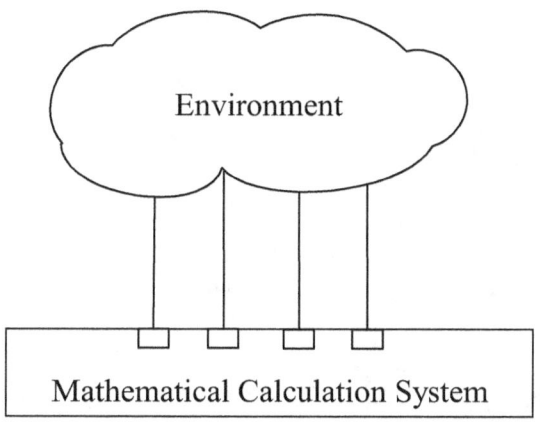

Figure 10-4　Systems Modeling of
the *Mathematical Calculation System*

10-1 BNF Tree of the Mathematical Calculation System

The channel-based single-queue SBC process of the *Mathematical Calculation System*, A_{031}, is defined as "$\mathbf{fix}(X_{031}=t_{031}\bullet t_{032}\bullet t_{033}\bullet t_{034}\bullet X_{031}+t_{035}\bullet t_{032}\bullet t_{037}\bullet t_{038}\bullet X_{031})$".

We draw the channel-based single-queue SBC process algebra Backus-Naur Form tree of A_{031} as shown in Figure 10-5.

81

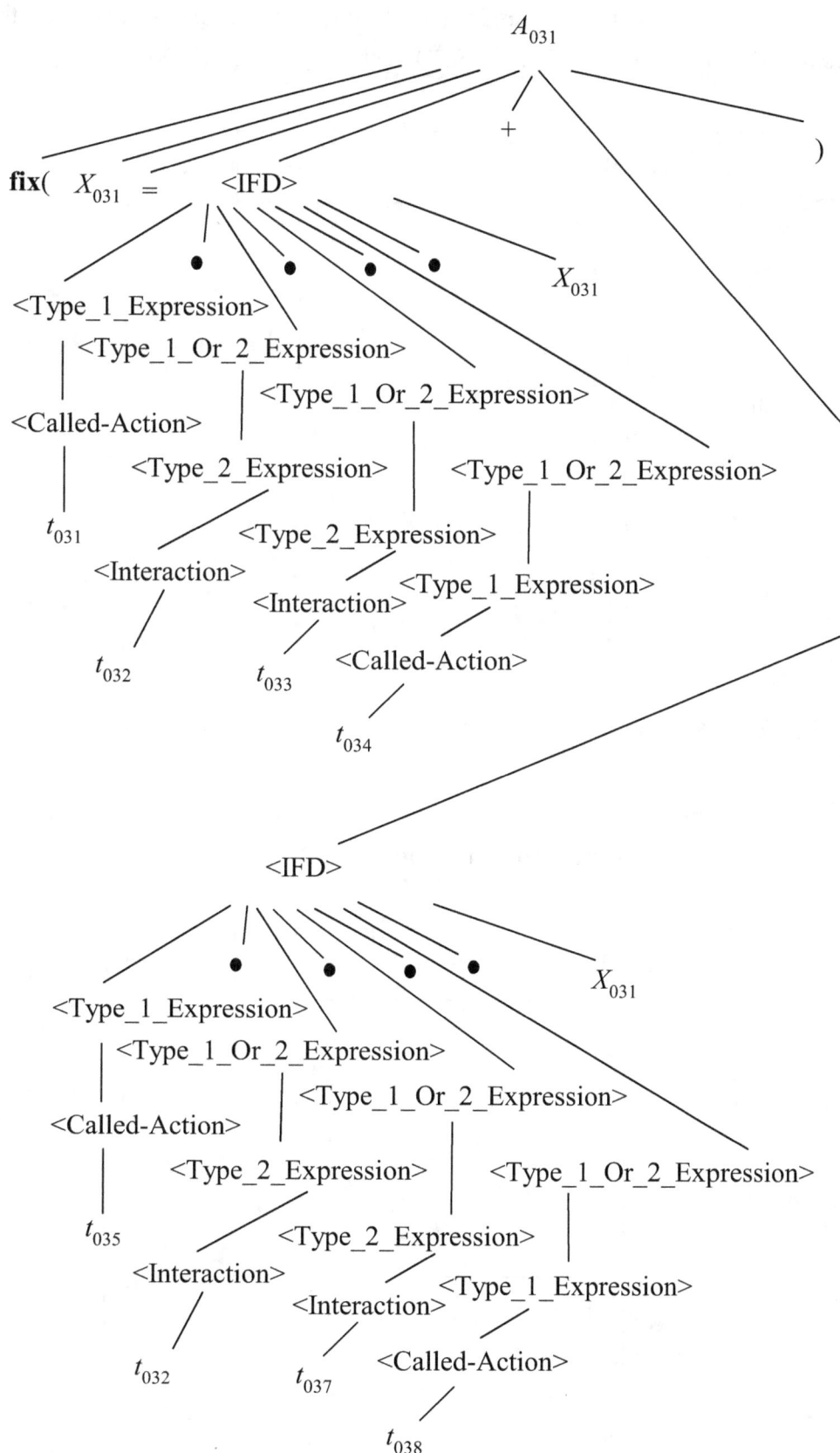

Figure 10-5   Backus-Naur Form Tree of the *Mathematical Calculation System*'s Channel-Based Single-Queue SBC Process

There are two IFDs in the channel-based single-queue SBC process of the *Mathematical Calculation System*. The first IFD is shown in Figure 10-6.

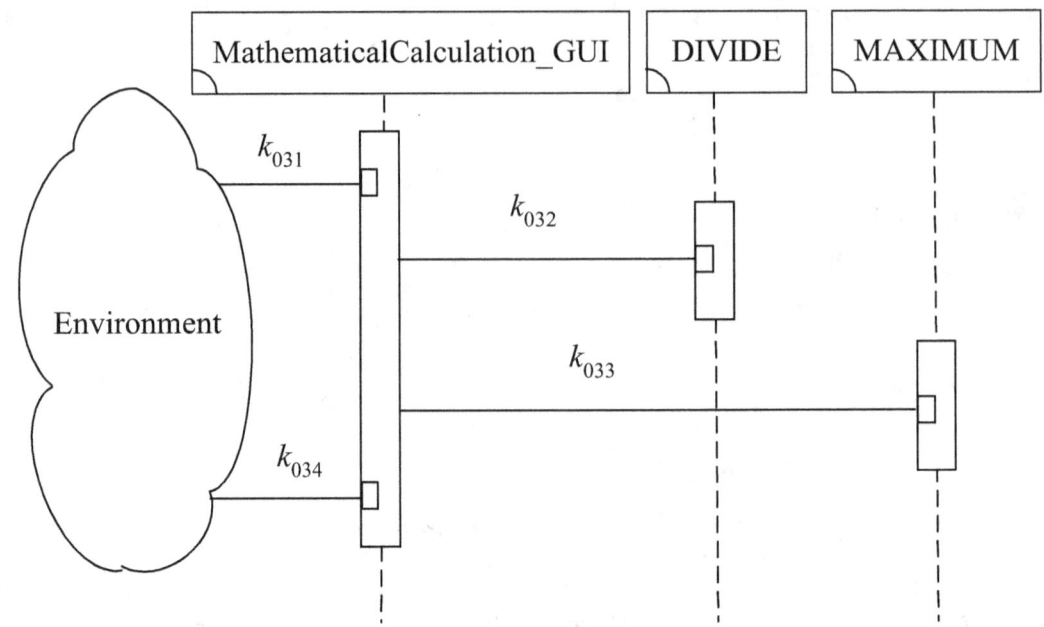

Figure 10-6 First IFD of the *Mathematical Calculation System*

The second IFD of the channel-based single-queue SBC process of the *Mathematical Calculation System* is shown in Figure 10-7.

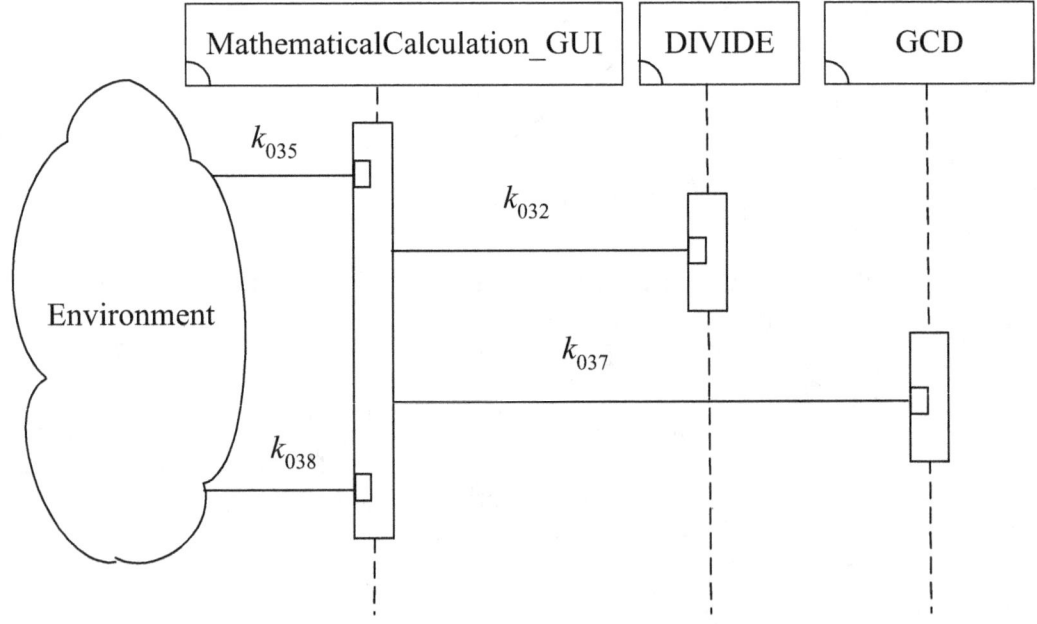

Figure 10-7 Second IFD of the *Mathematical Calculation System*

10-2 Prefixes of the Mathematical Calculation System

The t_{031} (channel-based called action with no conditions and no variable settings) prefix is defined as "<*MathematicalCalculation_GUI*, CALLED, k_{031}>", as shown in Figure 10-8.

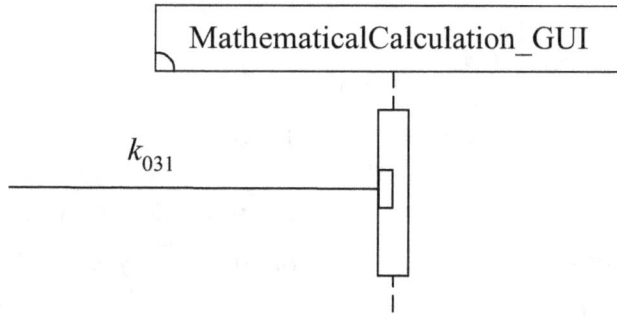

Figure 10-8 Prefix of t_{031}

The t_{032} (channel-based interaction with no conditions and no variable settings) prefix is defined as "<*MathematicalCalculation_GUI*, k_{032}, *DIVIDE*>", as shown in Figure 10-9.

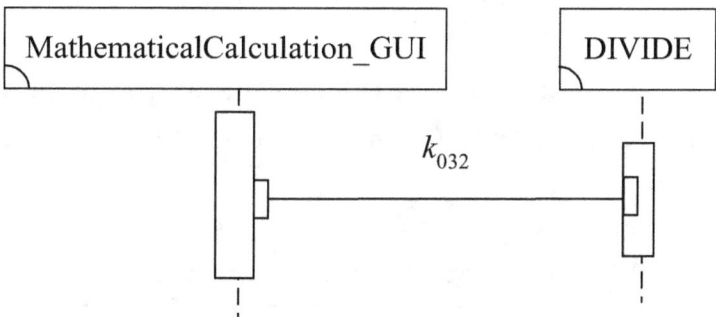

Figure 10-9 Prefix of t_{032}

The t_{033} (channel-based interaction with no conditions and no variable settings) prefix is defined as "<*DIVIDE*, k_{033}, *MAXIMUM*>", as shown in Figure 10-10.

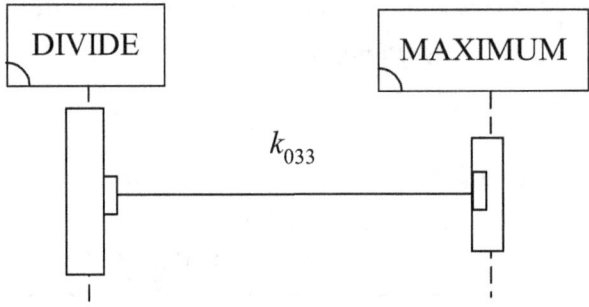

Figure 10-10   Prefix of t_{033}

The t_{034} (channel-based called action with no conditions and no variable settings) prefix is defined as "<*MathematicalCalculation_GUI*, CALLED, k_{034}>", as shown in Figure 10-11.

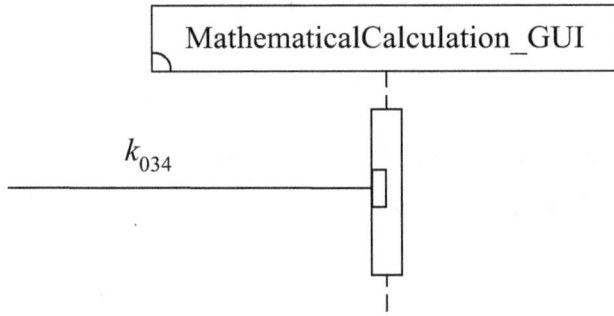

Figure 10-11   Prefix of t_{034}

The t_{035} (channel-based called action with no conditions and no variable settings) prefix is defined as "<*MathematicalCalculation_GUI*, CALLED, k_{035}>", as shown in Figure 10-12.

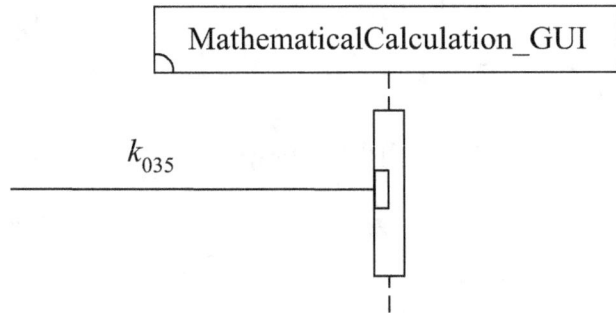

Figure 10-12   Prefix of t_{035}

The t_{037} (channel-based interaction with no conditions and no variable settings) prefix is defined as "<DIVIDE, k_{037}, GCD>", as shown in Figure 10-13.

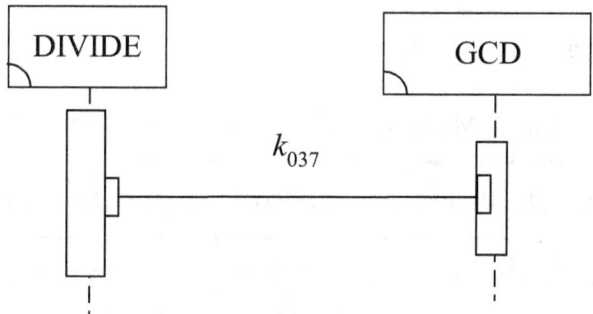

Figure 10-13 Prefix of t_{037}

The t_{038} (channel-based called action with no conditions and no variable settings) prefix is defined as "<MathematicalCalculation_GUI, CALLED, k_{038}>", as shown in Figure 10-14.

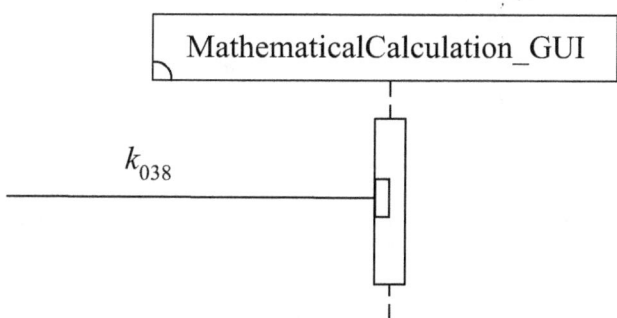

Figure 10-14 Prefix of t_{038}

Figure 10-15 shows all channel formulas of the channel-based single-queue SBC process of the *Mathematical Calculation System*.

Entity name	Channel Formula
k_{031}	Calculate_MaximumClick_Call(In Dividend, Divisor)
k_{032}	DIVIDE(In dividend, divisor; Out quotient, remainder)
k_{033}	MAXIMUM(In quotient, remainder; Out maximum_value)
k_{034}	Calculate_MaximumClick_Return(Out Maximum_value)
k_{035}	Calculate_GCDClick_Call(In Dividend, Divisor)
k_{037}	GCD(In quotient, remainder; Out gcd_value)
k_{038}	Calculate_GCDClick_Return(Out GCD_value)

Figure 10-15   Channel Formulas
of the *Mathematical Calculation System*'s Process

Figure 10-16 shows the primitive data type specification of the *Dividend*, *Divisor*, *Maximum_value*, *GCD_value*, *dividend*, *divisor*, *quotient*, *remain*, *maximum_value* and *gcd_value* parameters.

Parameter	Data Type	Instances
Dividend	Integer	88, 188
Divisor	Integer	5, 15
Maximum_value	Integer	17, 12
GCD_value	Integer	1, 4
dividend	Integer	88, 188
divisor	Integer	5, 15
quotient	Integer	17, 12
remain	Integer	3, 8
maximum_value	Integer	17, 12
gcd_value	Integer	1, 4

Figure 10-16   Primitive Data Type Specification

10-3 Process of the Mathematical Calculation System

The following transition graph shows, in Figure 10-17, the semantics of A_{031}'s channel-based single-queue SBC process.

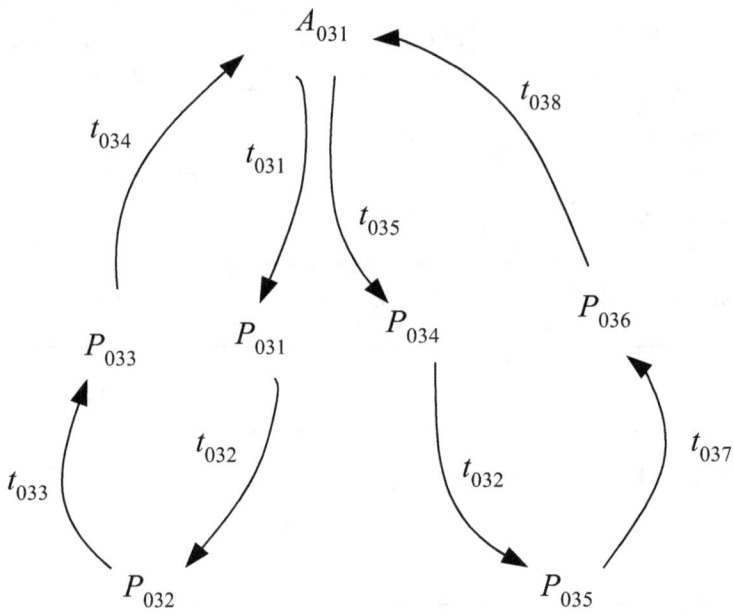

Figure 10-17   Transition graph
of the *Mathematical Calculation System*'s Process

In the transition graph of the A_{031}'s channel-based single-queue SBC process, processes A_{031}, P_{031}, P_{032}, P_{033}, P_{034}, P_{035} and P_{036} are defined as in Figure 10-18.

$$A_{031} \stackrel{\text{def}}{=} t_{031} \bullet P_{031} + t_{035} \bullet P_{034}$$

$$P_{031} \stackrel{\text{def}}{=} t_{032} \bullet P_{032}$$

$$P_{032} \stackrel{\text{def}}{=} t_{033} \bullet P_{033}$$

$$P_{033} \stackrel{\text{def}}{=} t_{034} \bullet A_{031}$$

$$P_{034} \stackrel{\text{def}}{=} t_{032} \bullet P_{035}$$

$$P_{035} \stackrel{\text{def}}{=} t_{037} \bullet P_{036}$$

$$P_{036} \stackrel{\text{def}}{=} t_{038} \bullet A_{031}$$

Figure 10-18 Definition of Processes A_{031}, P_{031}, P_{032}, P_{033}, P_{034}, P_{035}, and P_{036}

90

Chapter 11: Channel-Based Single-Queue SBC Process of the Web Service Arithmetic System

After the system development is completed, the *Web Service Arithmetic System* shall appear on a web multi-tier platform [Sebe12] as shown in Figure 11-1.

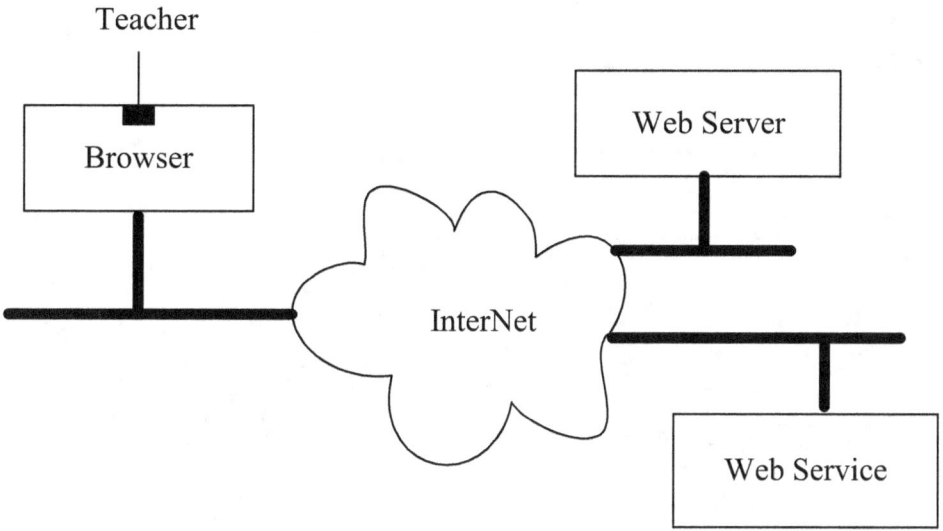

Figure 11-1 *Web Service Arithmetic System* on a Web Multi-Tier Platform

The functionality of the *Web Service Arithmetic System* is to provide a web browser for the *Teacher* actor to input the P, Q, R, S and T values as shown in Figure 11-2.

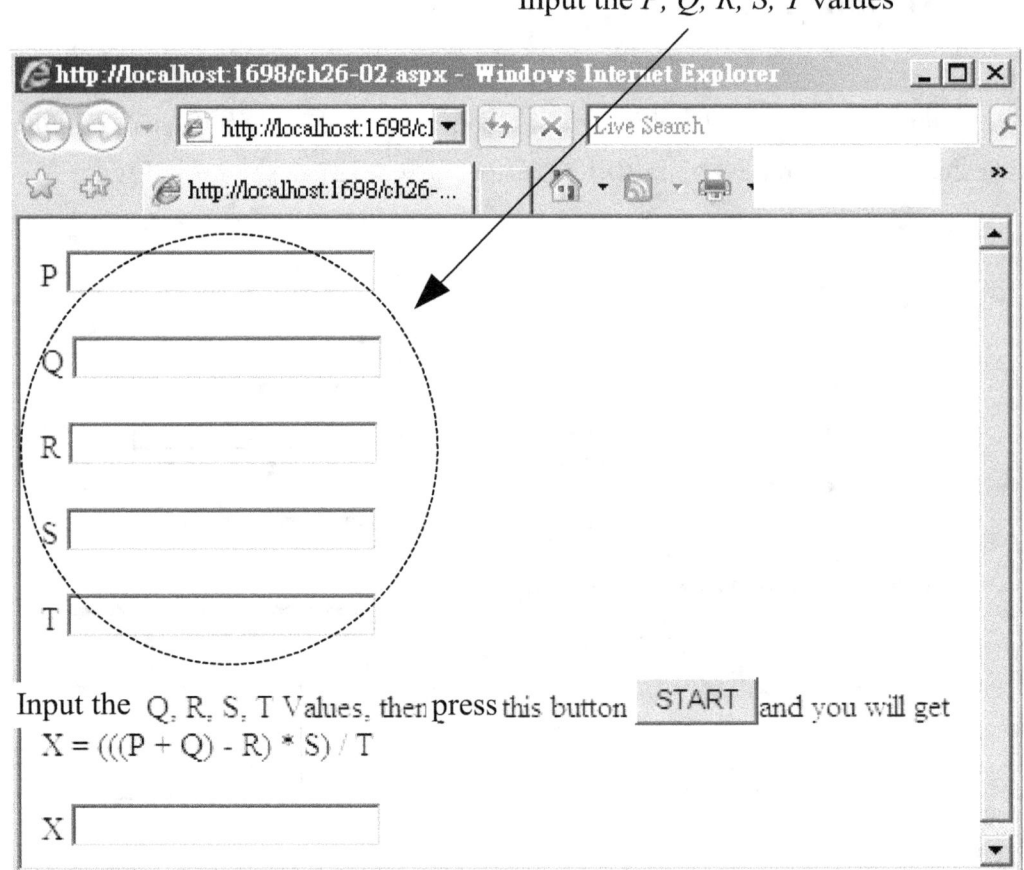

Figure 11-2   Web Browser for the *Pupil* Actor to Input the Values

When the *START* button is pressed, the *Web Service Arithmetic System* calculates the (((P+Q)-R) *S)/T value and presents the result on the *X* output box, as shown in Figure 11-3.

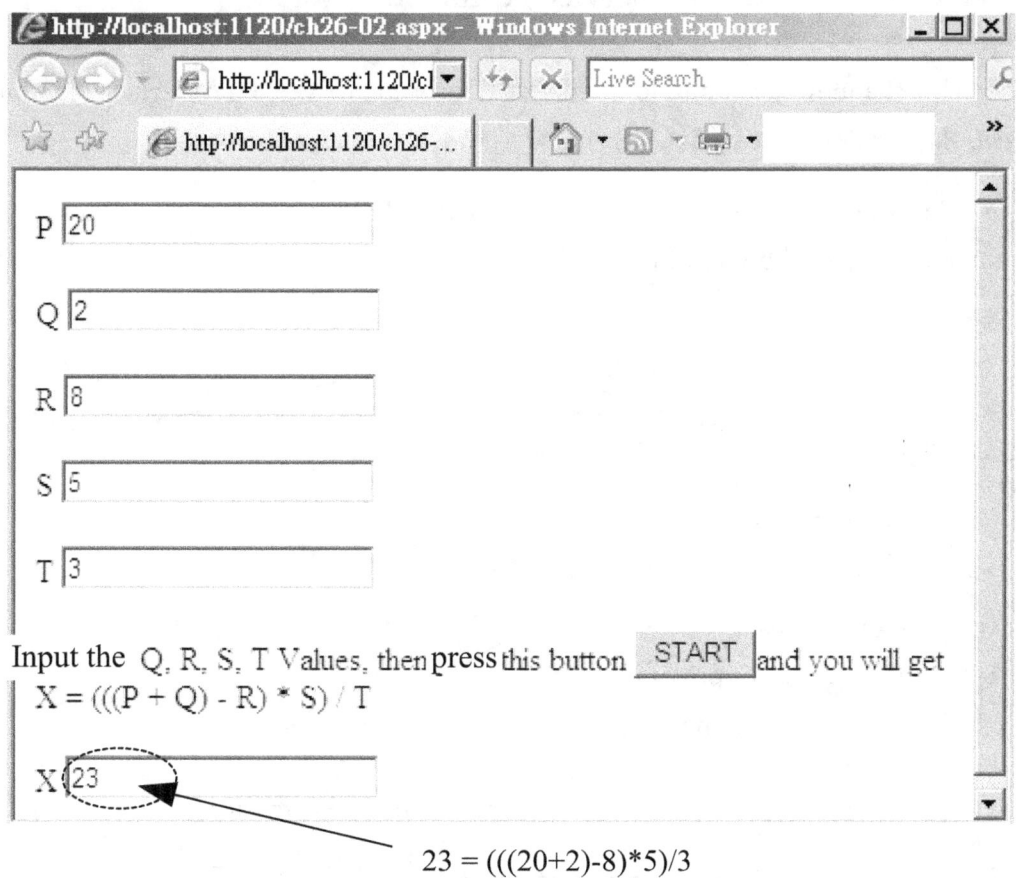

Figure 11-3 Result after Pressing Down the *START* Button

In this chapter, we use the channel-based single-queue SBC process algebra to model the *Web Service Arithmetic System* as shown in Figure 11-4.

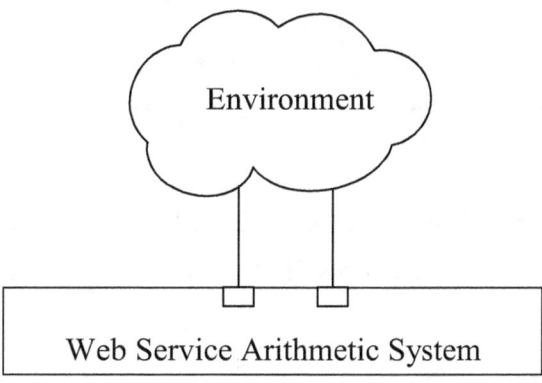

Figure 11-4 Systems Modeling
the *Web Service Arithmetic System*

11-1 BNF Tree of the Web Service Arithmetic System

The channel-based single-queue SBC process of the *Web Service Arithmetic System*, A_{041}, is defined as "**fix**$(X_{041}=t_{041} \bullet t_{042} \bullet t_{043} \bullet t_{044} \bullet t_{045} \bullet t_{046} \bullet X_{041})$".

We draw the channel-based single-queue SBC process algebra Backus-Naur Form tree of A_{041} as shown in Figure 11-5.

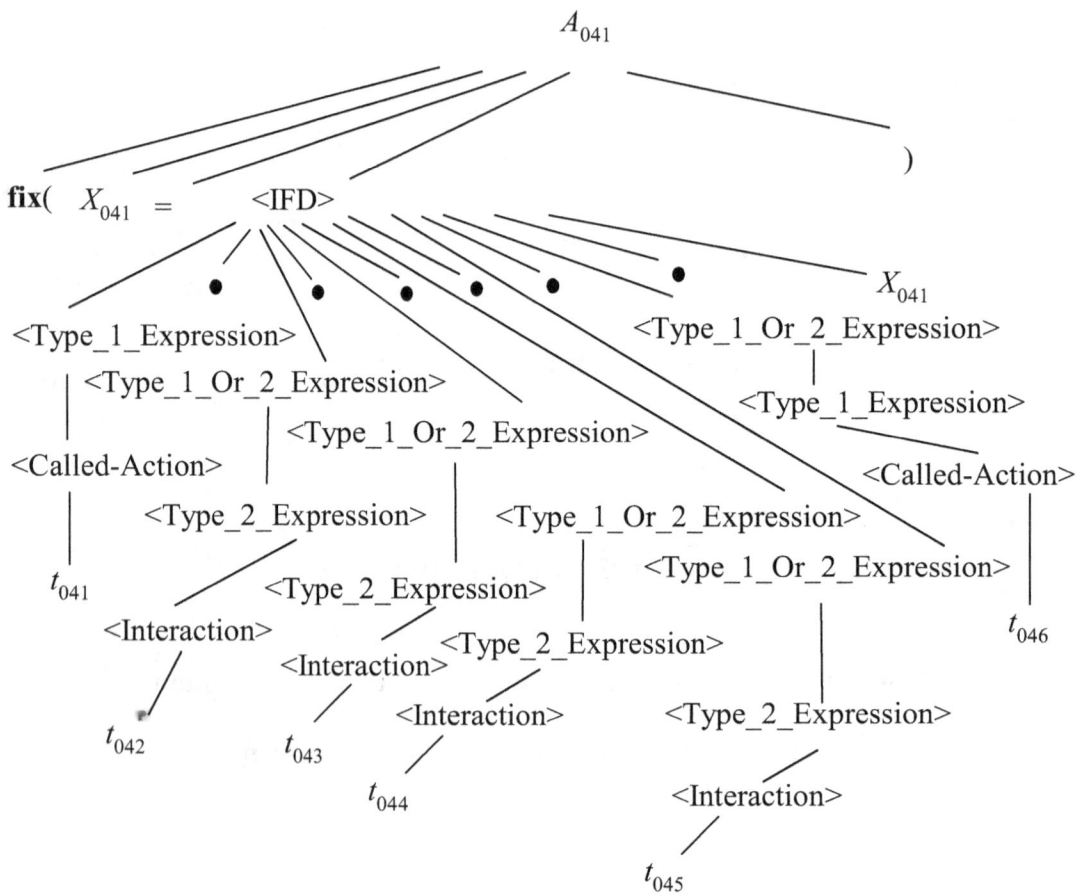

Figure 11-5　Backus-Naur Form Tree of the *Web Service Arithmetic System*'s Channel-Based Single-Queue SBC Process

There is only one IFD in the channel-based single-queue SBC process of the *Web Service Arithmetic System*. The only IFD is shown in Figure 11-6.

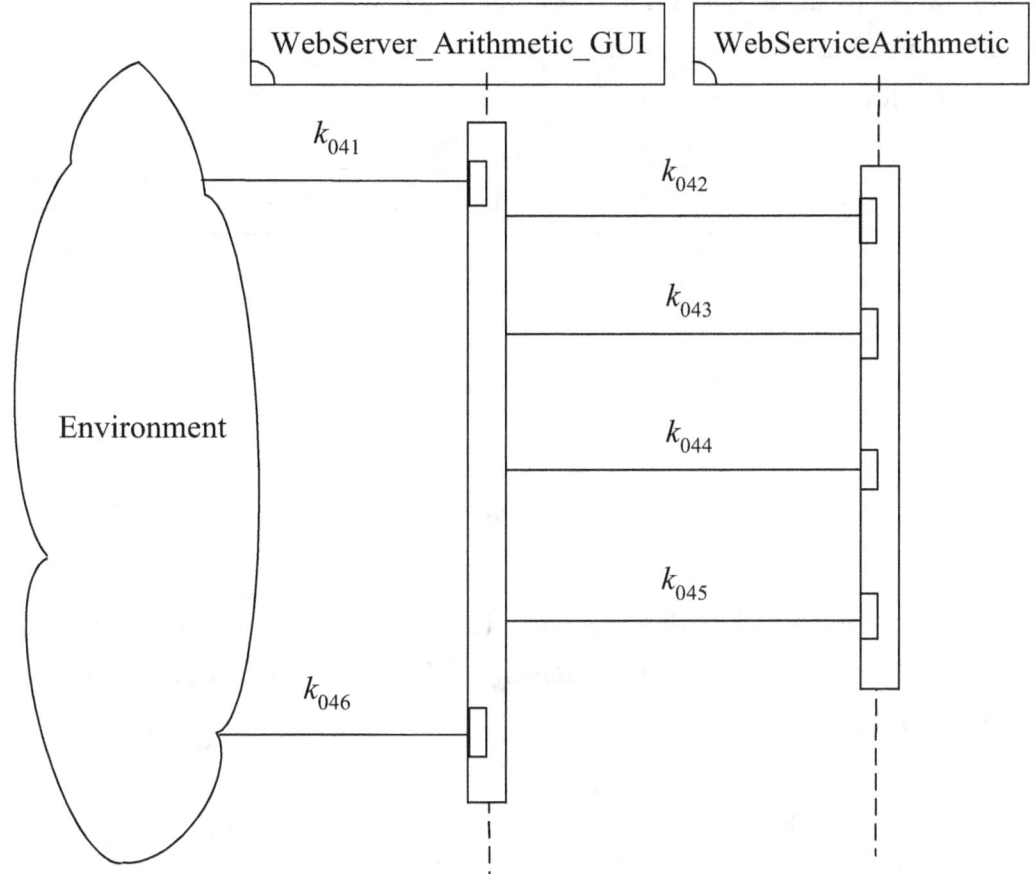

Figure 11-6   The Only IFD of the *Web Service Arithmetic System*

11-2 Prefixes of the Web Service Arithmetic System

The t_{041} (channel-based called action with no conditions and no variable settings) prefix is defined as "<*WebServer_Arithmetic_GUI*, CALLED, k_{041}>", as shown in Figure 11-7.

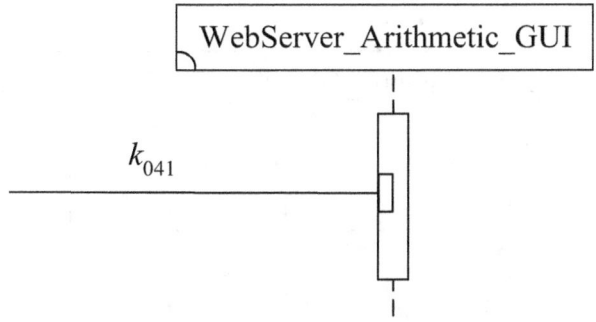

Figure 11-7   Prefix of t_{041}

The t_{042} (channel-based interaction with no conditions and no variable settings) prefix is defined as "<*WebServer_Arithmetic_GUI*, k_{042}, *WebServiceArithmetic*>", as shown in Figure 11-8.

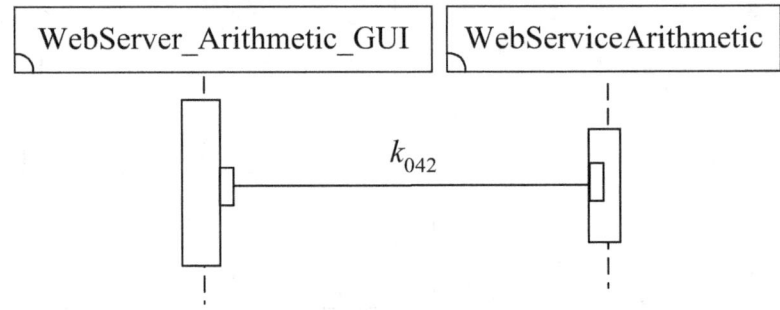

Figure 11-8   Prefix of t_{042}

The t_{043} (channel-based interaction with no conditions and no variable settings) prefix is defined as "<*WebServer_Arithmetic_GUI*, k_{043}, *WebServiceArithmetic*>", as shown in Figure 11-9.

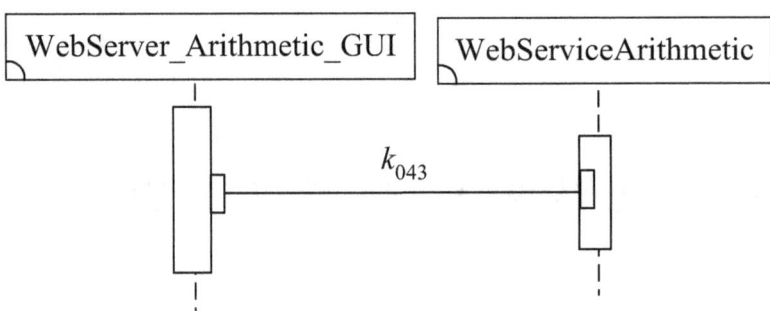

Figure 11-9   Prefix of t_{043}

The t_{044} (channel-based interaction with no conditions and no variable settings) prefix is defined as "<*WebServer_Arithmetic_GUI*, k_{044}, *WebServiceArithmetic*>", as shown in Figure 11-10.

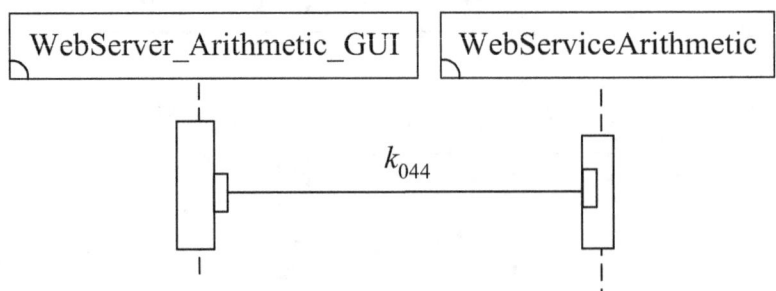

Figure 11-10   Prefix of t_{044}

The t_{045} (channel-based interaction with no conditions and no variable settings) prefix is defined as "<*WebServer_Arithmetic_GUI*, k_{045}, *WebServiceArithmetic*>", as shown in Figure 11-11.

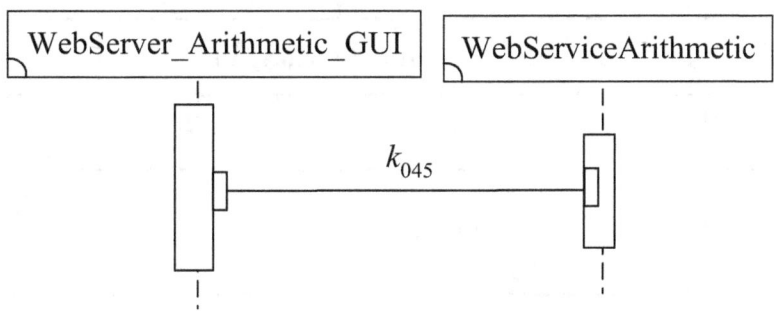

Figure 11-11   Prefix of t_{045}

The t_{046} (channel-based called action with no conditions and no variable settings) prefix is defined as "<*WebServer_Arithmetic_GUI*, CALLED, k_{046}>", as shown in Figure 11-12.

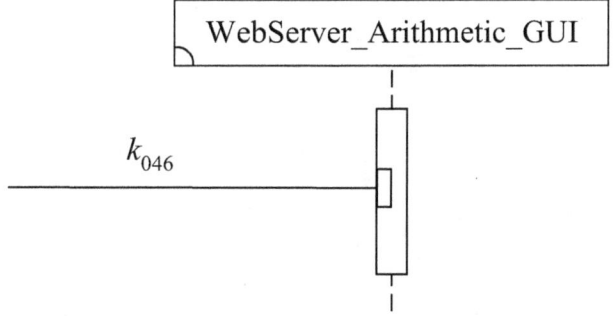

Figure 11-12   Prefix of t_{046}

Figure 11-13 shows all channel formulas of the channel-based single-queue SBC process of the *Web Service Arithmetic System*.

Entity name	Channel Formula
k_{041}	START_Click_Call(In P, Q, R, S, T)
k_{042}	Add(In a, b; Out c)
k_{043}	Subtract(In d, e; Out f)
k_{044}	Multiply(In g, h; Out i)
k_{045}	Divide(In j, k; Out m)
k_{046}	START_Click_Return(Out X)

Figure 11-13 Channel Formulas
of the *Web Service Arithmetic System*'s Process

Figure 11-14 shows the primitive data type specification of the *P, Q, R, S, T, a, b, d, e, g, h, j, k* input parameter and the *X, c, f, i, m* output parameters.

Parameter	Data Type	Instances
P	Integer	20, 7
Q	Integer	2, 6
R	Integer	8, 5
S	Integer	5, 4
T	Integer	3, 3
X	Integer	23, 11
a	Integer	20, 7
b	Integer	2, 6
c	Integer	22, 13
d	Integer	22, 13
e	Integer	8, 5
f	Integer	14, 8
g	Integer	14, 8
h	Integer	5, 4
i	Integer	70, 32
j	Integer	70, 32
k	Integer	3, 3
m	Integer	23, 11

Figure 11-14   Primitive Data Type Specification

11-3 Process of the Web Service Arithmetic System

The following transition graph shows, in Figure 11-15, the semantics of A_{041}'s channel-based single-queue SBC process.

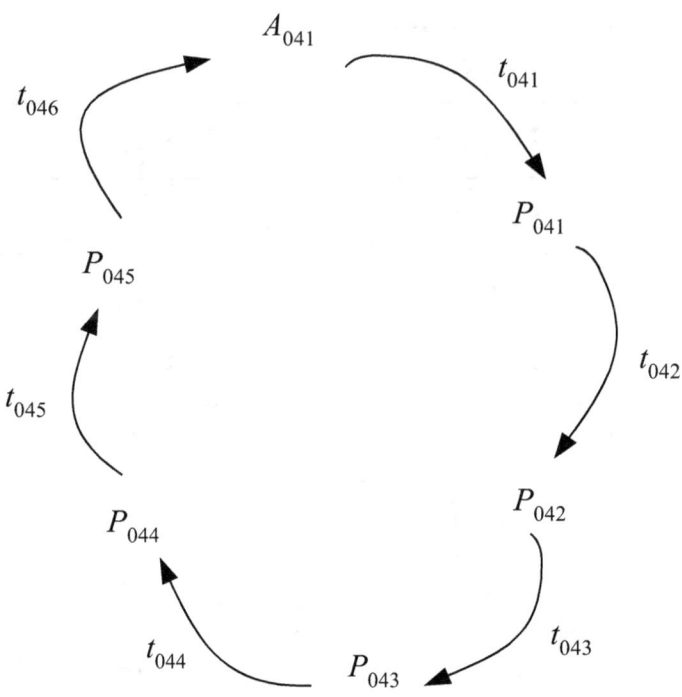

Figure 11-15   Transition graph
of the *Web Service Arithmetic System*'s Process

In the transition graph of the A_{041}'s channel-based single-queue SBC process, processes $A_{041}, P_{041}, P_{042}, P_{043}, P_{044}$ and P_{045} are defined as in Figure 11-16.

$$A_{041} \stackrel{\text{def}}{=\!=} t_{041} \bullet P_{041}$$

$$P_{041} \stackrel{\text{def}}{=\!=} t_{042} \bullet P_{042}$$

$$P_{042} \stackrel{\text{def}}{=\!=} t_{043} \bullet P_{043}$$

$$P_{043} \stackrel{\text{def}}{=\!=} t_{044} \bullet P_{044}$$

$$P_{044} \stackrel{\text{def}}{=\!=} t_{045} \bullet P_{045}$$

$$P_{045} \stackrel{\text{def}}{=\!=} t_{046} \bullet A_{041}$$

Figure 11-16 Definition of Processes $A_{041}, P_{041}, P_{042}, P_{043}, P_{044},$ and P_{045}

Chapter 12: Channel-Based Single-Queue SBC Process of the Web Service Extranet System

After the system development is completed, the *web service extranet system* shall appear on a web multi-tier platform [Sebe12] as shown in Figure 12-1.

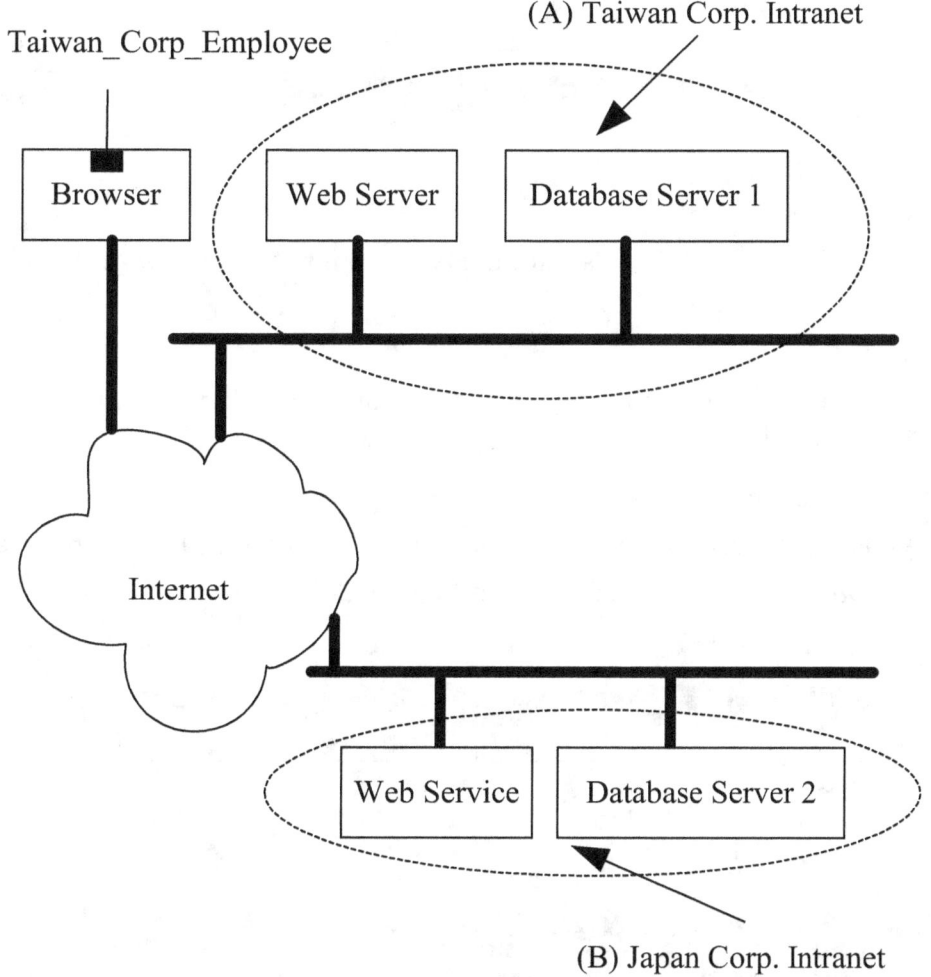

Figure 12-1 *Web Service Extranet System* on a Web Multi-Tier Platform

The major functionality of the *Web Service Extranet System* is to provide a web browser for the *Taiwan_Corp_Employee* actor to input the *PurchaseDate* value as shown in Figure 12-2.

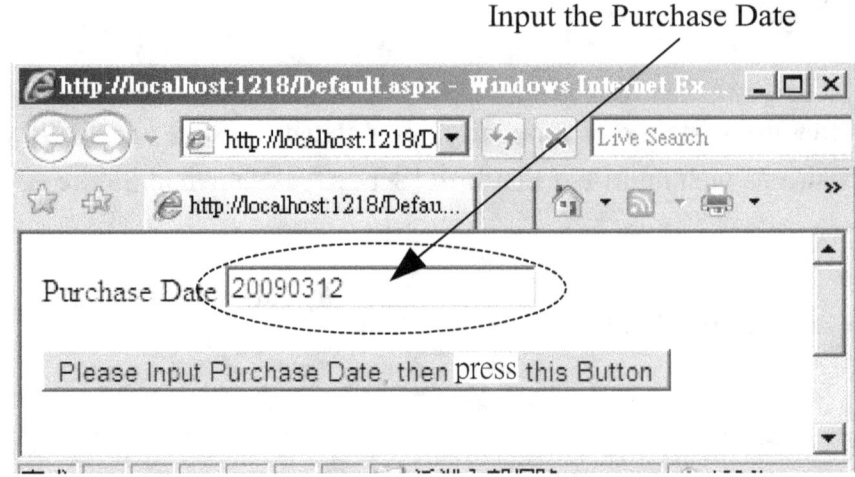

Figure 12-2   Web Browser for the Actor to input the *Purchase Date*

After the *Taiwan_Corp_Employee* actor inputs the *PurchaseDate* value and presses down the *ExtranetButton* button then the *Web Service Extranet System* generates the following two results:

(A) The first result is to insert a new record into the *PurchaseTable* table of the *Taiwan_Database* database as shown in Figure 12-3.

Figure 12-3   First Result After Pressing Down the *ExtranetButton* Button

(B) The second result is to insert a new record into the *SaleTable* table of the *Japan_Database* database as shown in Figure 12-4.

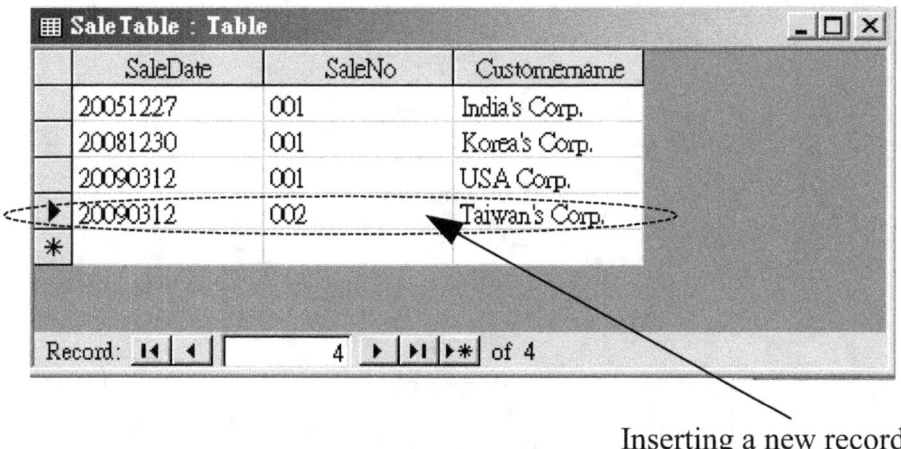

Inserting a new record

Figure 12-4   Second Result After Pressing Down the *ExtranetButton* Button

Finally, we also see a message informing us that the Extranet deal is done, on the web page, as shown in Figure 12-5.

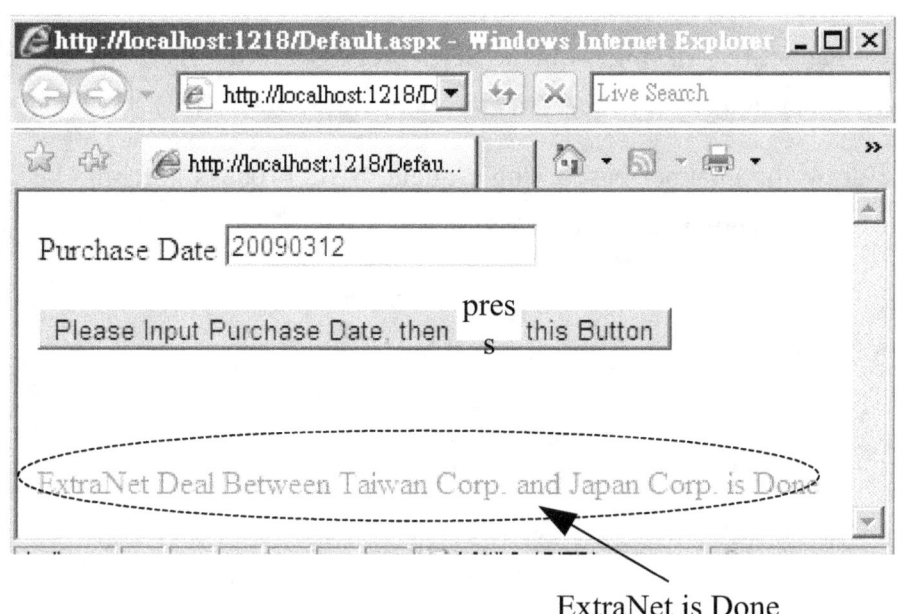

ExtraNet is Done

Figure 12-5   Informing that the Extranet Deal is Done

In this chapter, we use the channel-based single-queue SBC process algebra to model the *Web Service Extranet System* as shown in Figure 12-6.

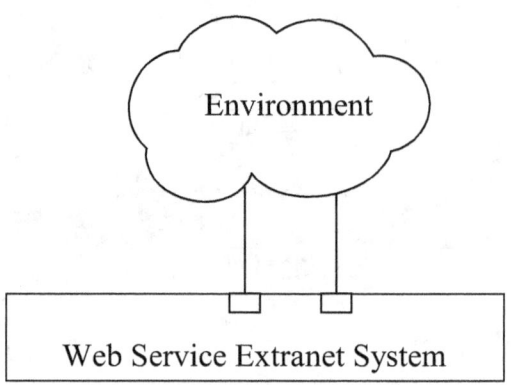

Figure 12-6   Systems Modeling
the *Web Service Extranet System*

12-1 BNF Tree of the Web Service Extranet System

The channel-based single-queue SBC process of the *Web Service Extranet System*, A_{081}, is defined as "$\mathbf{fix}(X_{081}=t_{081} \bullet t_{082} \bullet t_{083} \bullet t_{084} \bullet t_{085} \bullet X_{081})$".

We draw the channel-based single-queue SBC process algebra Backus-Naur Form tree of A_{081} as shown in Figure 12-7.

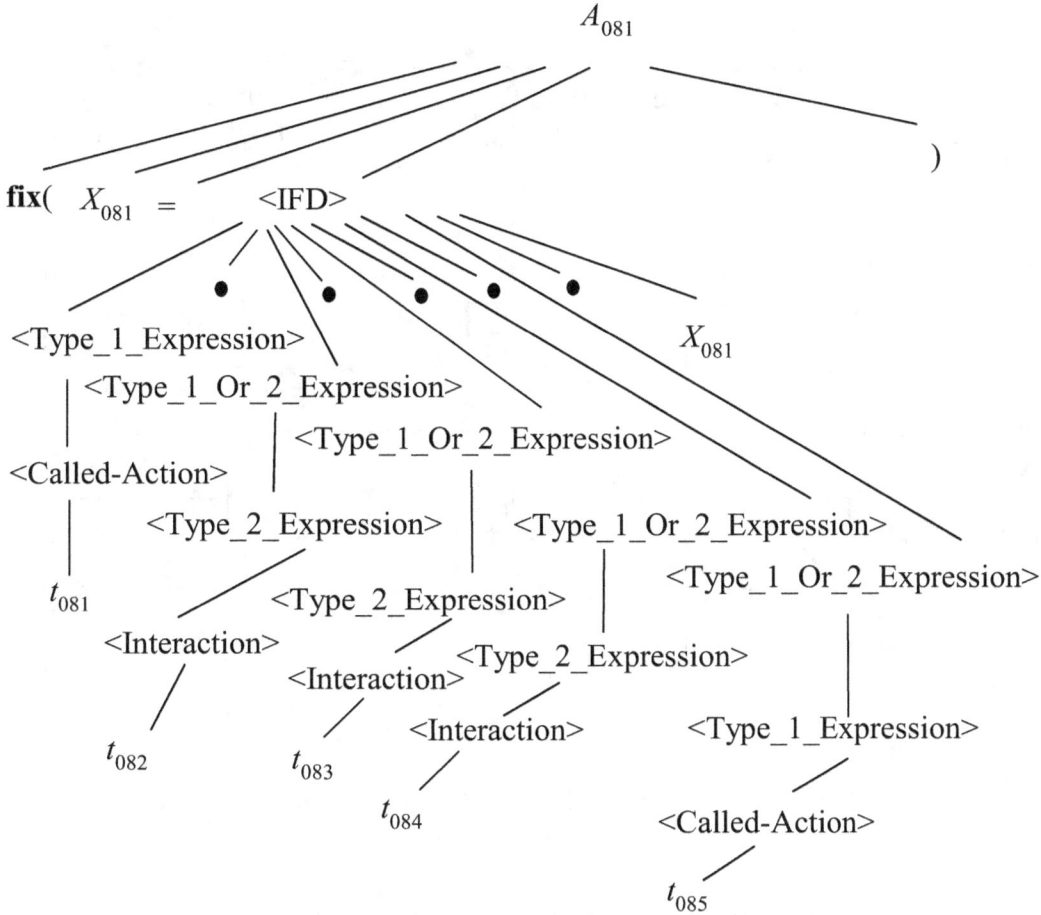

Figure 12-7 Backus-Naur Form Tree of the *Web Service Extranet System*'s Channel-Based Single-Queue SBC Process

There is only one IFD in the channel-based single-queue SBC process of the *Web Service Extranet System*. The only IFD is shown in Figure 12-8.

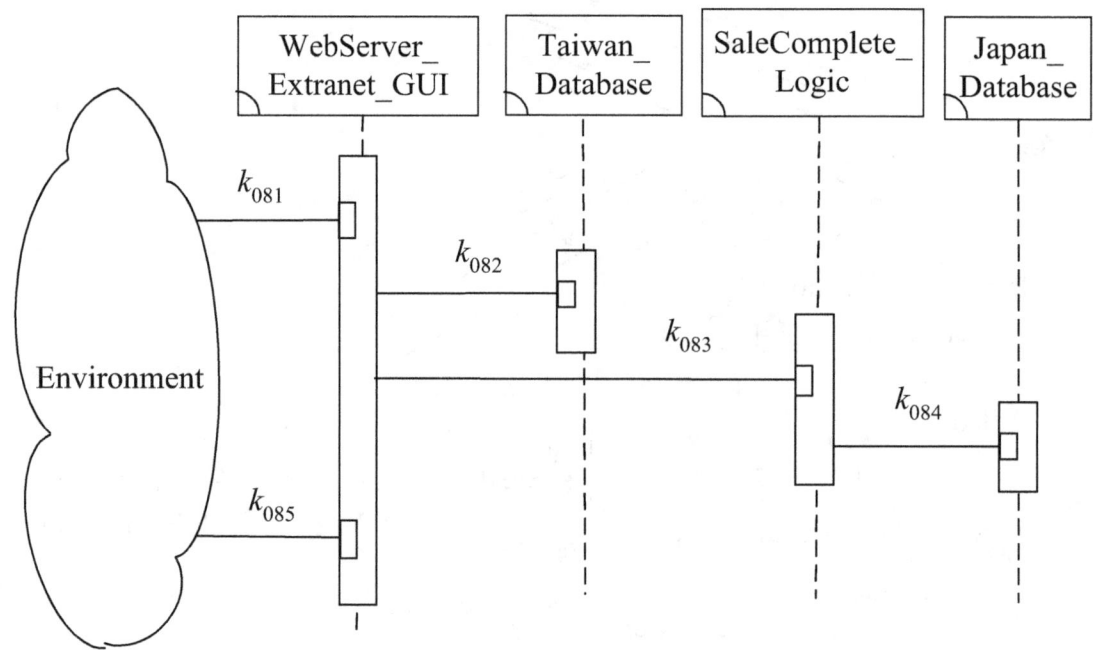

Figure 12-8 IFD of the *Web Service Extranet System*

12-2 Prefixes of the Web Service Extranet System

The t_{081} (channel-based called action with no conditions and no variable settings) prefix is defined as "<*WebServer_Extranet_GUI*, CALLED, k_{081}>", as shown in Figure 12-9.

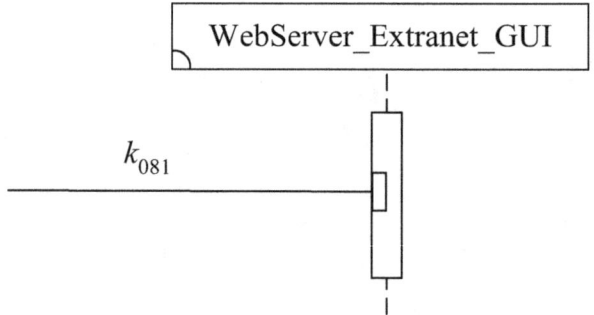

Figure 12-9 Prefix of t_{081}

The t_{082} (channel-based interaction with no conditions and no variable settings) prefix is defined as "<*WebServer_Extranet_GUI*, k_{082}, *Taiwan_Database*>", as shown in Figure 12-10.

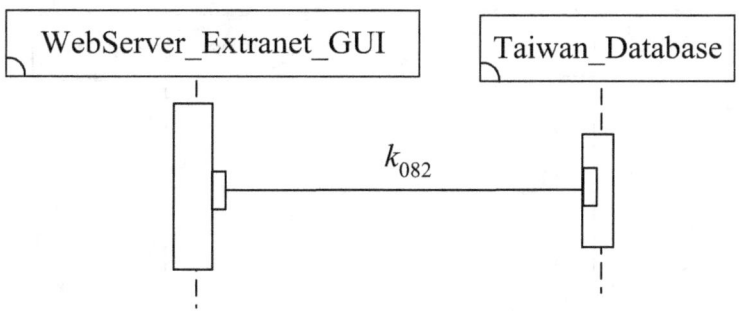

Figure 12-10　Prefix of t_{082}

The t_{083} (channel-based interaction with no conditions and no variable settings) prefix is defined as "<*WebServer_Extranet_GUI*, k_{083}, *SaleComplete_Logic*>", as shown in Figure 12-11.

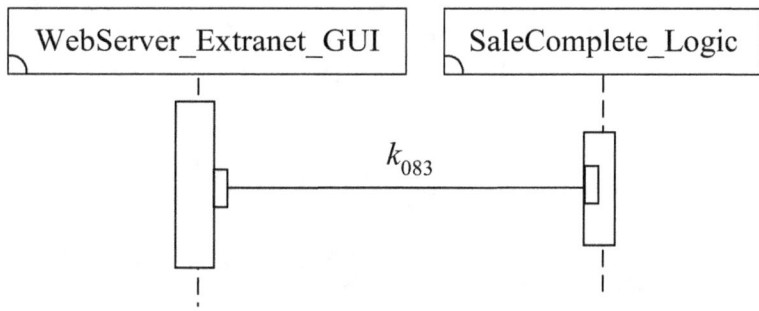

Figure 12-11　Prefix of t_{083}

The t_{084} (channel-based interaction with no conditions and no variable settings) prefix is defined as "<*SaleComplete_Logic*, k_{084}, *Japan_Database*>", as shown in Figure 12-12.

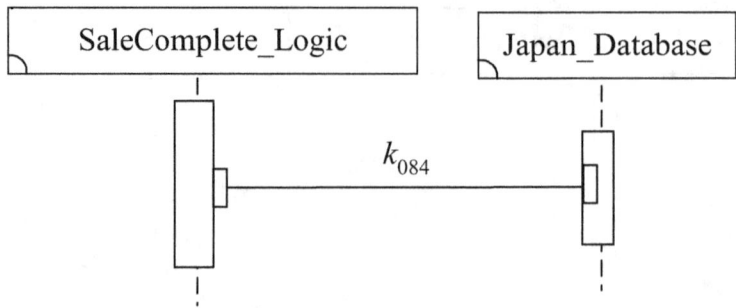

Figure 12-12　Prefix of t_{084}

The t_{085} (channel-based called action with no conditions and no variable settings) prefix is defined as "<*WebServer_Extranet_GUI*, CALLED, k_{085}>", as shown in Figure 12-13.

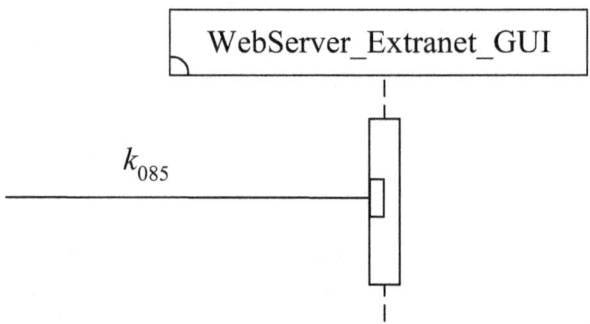

Figure 12-13　Prefix of t_{085}

Figure 12-14 shows all channel formulas of the channel-based single-queue SBC process of the *Web Service Extranet System*.

Entity name	Channel Formula
k_{081}	ExtranetButton_Click_Call(In PurchaseDate)
k_{082}	Add_a_Sale(In SaleDate, Customer)
k_{083}	Sql_p_Insert is Sql_p_Insert(In p_query)
k_{084}	Sql_s_Insert is Sql_s_Insert(In s_query)
k_{085}	ExtranetButton_Click_Return

Figure 12-14　Channel Formulas
of the *Web Service Extranet System*'s Process

Figure 12-15 shows the primitive data type specification of the *PurchaseDate*, *SaleDate* and *Customer* input parameters.

Parameter	Data Type	Instances
PurchaseDate	String	20090112
SaleDate	String	20090112
Customer	String	Taiwan Corp.

Figure 12-15　Primitive Data Type Specification

Figure 12-16 shows the composite data type specification of the *p_query* input parameter occurring in the *Sql_p_Insert(In p_query)* channel formula.

Parameter	*p_query*

Data Type	TABLE of
	   Purchase Date : Text
	   Purchase No : Text
	   Supplier : Text
	End TABLE ;

Instances	<table><tr><td>PurchaseDate</td><td>PurchaseNo</td><td>Supplier</td></tr><tr><td>20090312</td><td>001</td><td>Japan Corp.</td></tr></table>

Figure 12-16   Composite Data Type Specification

Figure 12-17 shows the composite data type specification of the *s_query* input parameter occurring in the *Sql_s_Insert(In s_query)* channel formula.

Parameter	*s_query*

Data Type	TABLE of
	   SaleDate : Text
	   SaleNo : Text
	   Customer : Text
	End TABLE ;

Instances	<table><tr><td>Sale Date</td><td>Sale No</td><td>Customer</td></tr><tr><td>20090312</td><td>001</td><td>Taiwan Corp.</td></tr></table>

Figure 12-17   Composite Data Type Specification

12-3 Process of the Web Service Extranet System

The following transition graph shows, in Figure 12-18, the semantics of A_{081}'s channel-based single-queue SBC process.

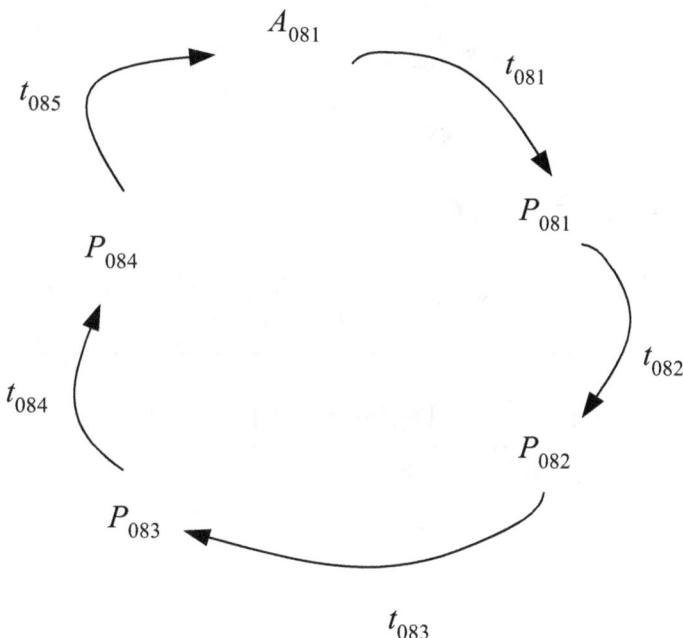

Figure 12-18 Transition graph of the *Web Service Extranet System*'s Process

In the transition graph of the A_{081}'s channel-based single-queue SBC process, processes A_{081}, P_{081}, P_{082}, P_{083} and P_{084} are defined as in Figure 12-19.

$$A_{081} \stackrel{def}{=} t_{081} \bullet P_{081}$$

$$P_{081} \stackrel{def}{=} t_{082} \bullet P_{082}$$

$$P_{082} \stackrel{def}{=} t_{083} \bullet P_{083}$$

$$P_{083} \stackrel{def}{=} t_{084} \bullet P_{084}$$

$$P_{084} \stackrel{def}{=} t_{085} \bullet A_{081}$$

Figure 12-19 Definition of Processes $A_{081}, P_{081}, P_{082}, P_{083}$, and P_{084}

Chapter 13: Channel-Based Single-Queue SBC Process of the Convenience Store's Get 2nd 50% off Sales Promotion System

In order to attract customers to buy, a convenience store may find many ways to do so. The easiest one is to give discounts. Because of human nature, coupled with the hot summer weather, the customers see the advertisement saying a cold drink at 50% off, they may have a great desire to buy it. This inspires the idea of a *Convenience Store's Get 2nd 50% off Sales Promotion System*.

In this chapter, we use the channel-based single-queue SBC process algebra to model the *Convenience Store's Get 2nd 50% off Sales Promotion System* as shown in Figure 13-1.

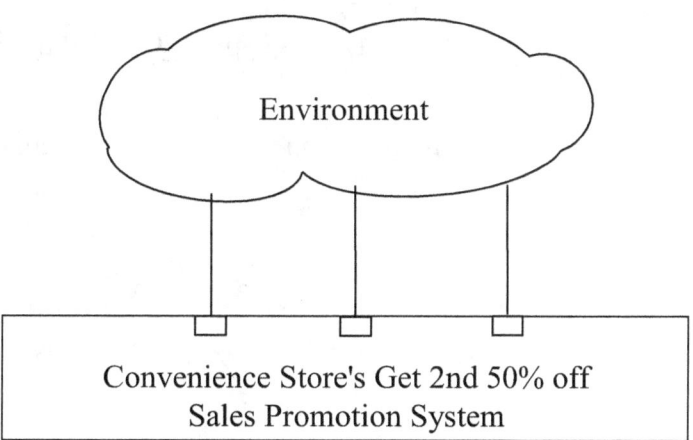

Figure 13-1 Systems Modeling the *Convenience Store's Get 2nd 50% off Sales Promotion System*

13-1 BNF Tree of the Convenience Store's Get 2nd 50% off Sales Promotion System

The channel-based single-queue SBC process of the *Convenience Store's Get 2nd 50% off Sales Promotion System*, A_{101}, is defined as "**fix**$(X_{101}=t_{101} \bullet X_{101}+t_{102} \bullet t_{103} \bullet X_{101})$".

We draw the channel-based single-queue SBC process algebra Backus-Naur Form tree of A_{101} as shown in Figure 13-2.

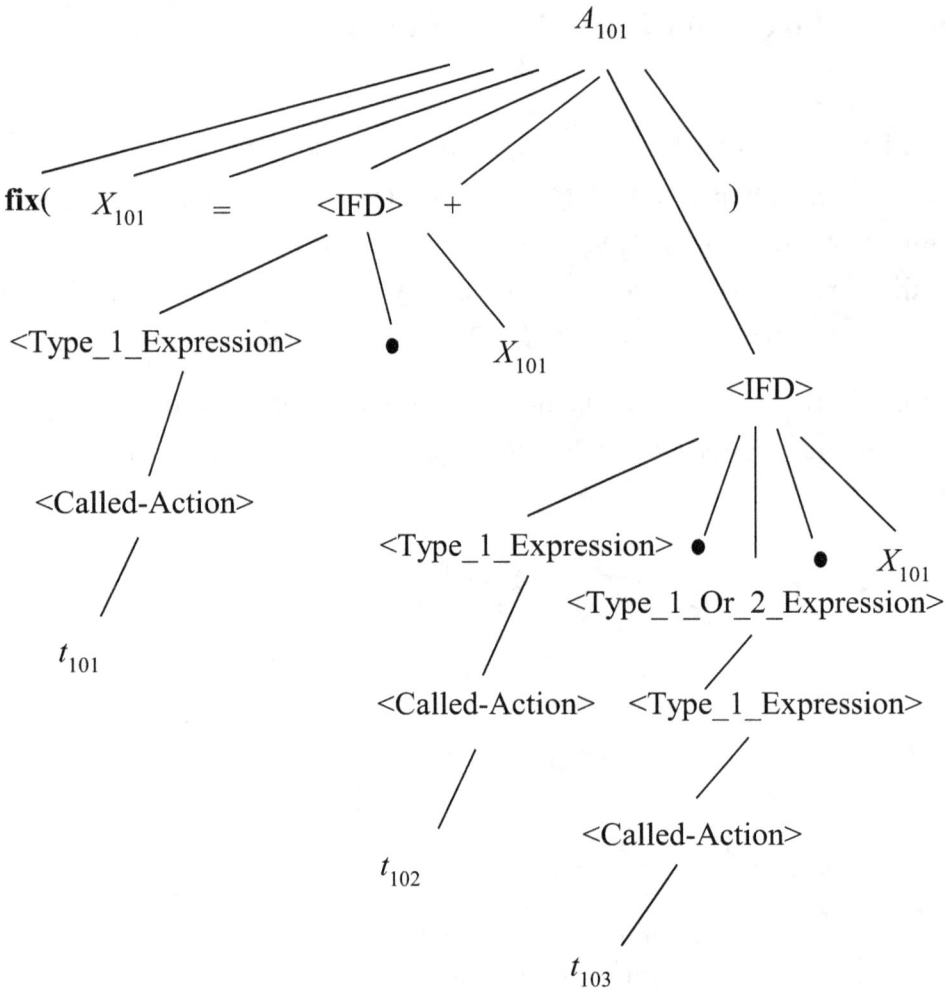

Figure 13-2 Backus-Naur Form Tree of the *Convenience Store's Get 2nd 50% off Sales Promotion System*'s Channel-Based Single-Queue SBC Process

There are two IFDs in the channel-based single-queue SBC process of the *Convenience Store's Get 2nd 50% off Sales Promotion System*. The first IFD is shown in Figure 13-3.

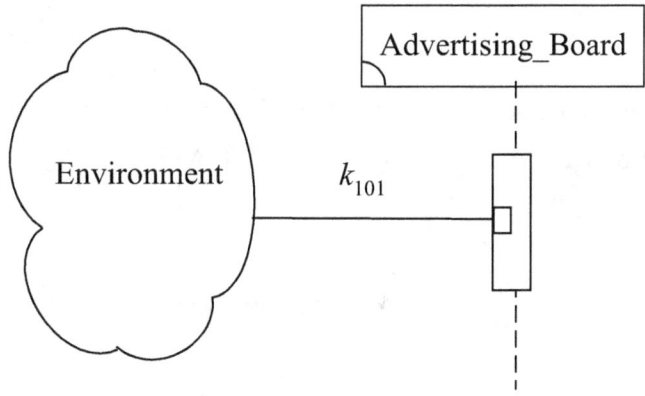

Figure 13-3 First IFD of the *Convenience Store's Get 2nd 50% off Sales Promotion System*

The second IFD of the channel-based single-queue SBC process of the *Convenience Store's Get 2nd 50% off Sales Promotion System* is shown in Figure 13-4.

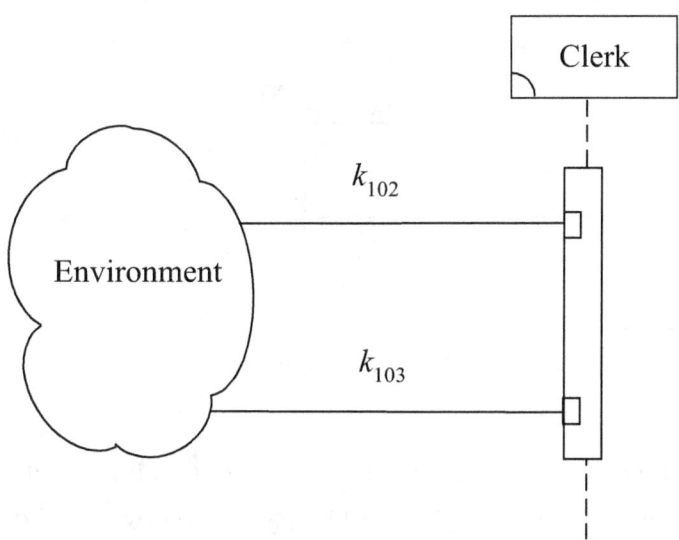

Figure 13-4 Second IFD of the *Convenience Store's Get 2nd 50% off Sales Promotion System*

13-2 Prefixes of the Convenience Store's Get 2nd 50% off Sales Promotion System

The t_{101} (channel-based called action with no conditions and no variable settings) prefix is defined as "<*Advertising_Board*, CALLED, k_{101}>", as shown in Figure 13-5.

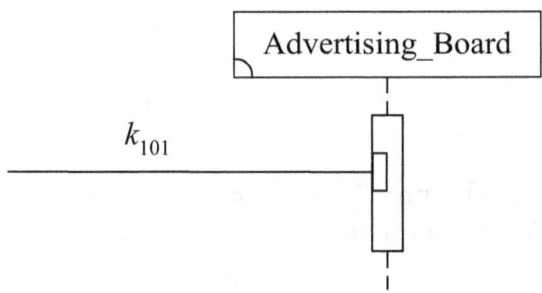

Figure 13-5 Prefix of t_{101}

The t_{102} (channel-based called action with no conditions and no variable settings) prefix is defined as "<*Clerk*, CALLED, k_{102}>", as shown in Figure 13-6.

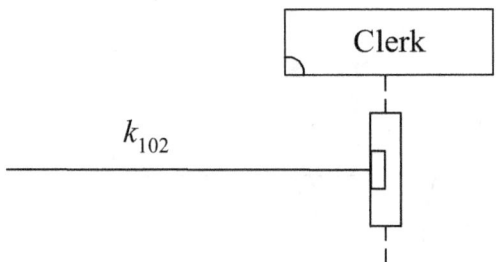

Figure 13-6 Prefix of t_{102}

The t_{103} (channel-based called action with no conditions and no variable settings) prefix is defined as "<*Clerk*, CALLED, k_{103}>", as shown in Figure 13-7.

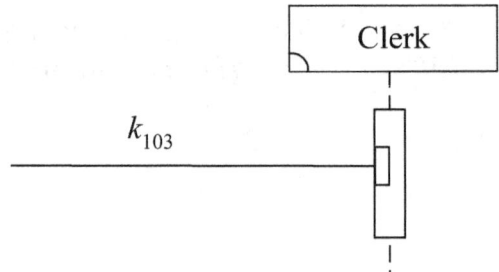

Figure 13-7 Prefix of t_{103}

Figure 13-8 shows all channel formulas of the channel-based single-queue SBC process of the *Convenience Store's Get 2nd 50% off Sales Promotion System.*

Entity name	Channel Formula
k_{101}	Watch(Out Advertisement)
k_{102}	Buy_the_1st_Goods(In Payment; Out Goods)
k_{103}	Buy_the_2nd_Goods(In 50%_off_Payment; Out Cold_Drink)

Figure 13-8   Channel Formulas of the *Convenience Store's Get 2nd 50% off Sales Promotion System*'s Process

Figure 13-9 shows the primitive data type specification of the *Payment, 50%_off_Payment* input parameters and the *Advertisement, Goods, Cold_Drink* output parameters.

Parameter	Data Type	Instances
Payment	Money	$100
50%_off_Payment	Money	$50
Advertisement	Text	Welcome here!!!
Goods	Bulk	3, 9
Cold_Drink	Bottle	20, 30

Figure 13-9   Primitive Data Type Specification

13-3 Process of the Convenience Store's Get 2nd 50% off Sales Promotion System

The following transition graph shows, in Figure 13-10, the semantics of A_{101}'s channel-based single-queue SBC process.

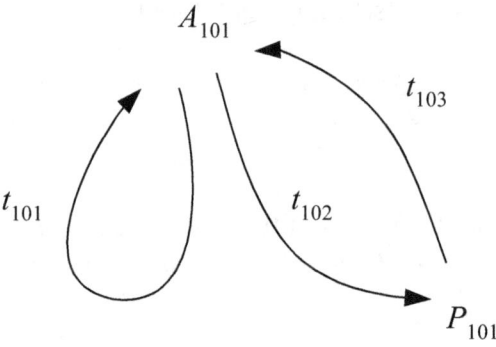

Figure 13-10 Transition graph of the *Convenience Store's Get 2nd 50% off Sales Promotion System*'s Process

In the transition graph of the A_{101}'s channel-based single-queue SBC process, processes A_{101} and P_{101} are defined as in Figure 13-11.

$$A_{101} \stackrel{\text{def}}{=} t_{101} \bullet A_{101} + t_{102} \bullet P_{101}$$
$$P_{101} \stackrel{\text{def}}{=} t_{103} \bullet A_{101}$$

Figure 13-11 Definition of Processes A_{101} and P_{101}

Chapter 14: Channel-Based Single-Queue SBC Process of the Department Store's Car Sweepstakes Sales Promotion System

A decent car is something everyone wants to have. A customer will be interested in buying a thousand dollars goods in a department store if he has the opportunity to get a grand new one-million-dollars car. This inspires the idea of a *Department Store's Car Sweepstakes Sales Promotion System.*

In this chapter, we use the channel-based single-queue SBC process algebra to model the *Department Store's Car Sweepstakes Sales Promotion System* as shown in Figure 14-1.

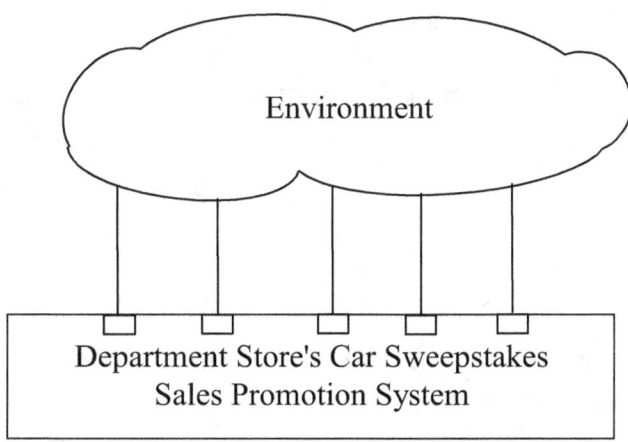

Figure 14-1 Systems Modeling the *Department Store's Car Sweepstakes Sales Promotion System*

14-1 BNF Tree of the Department Store's Car Sweepstakes Sales Promotion System

The channel-based single-queue SBC process of the *Department Store's Car Sweepstakes Sales Promotion System*, A_{141}, is defined as "**fix**$(X_{141}=t_{141} \bullet t_{142} \bullet X_{141}+t_{143} \bullet t_{144} \bullet t_{145} \bullet X_{141}+ t_{146} \bullet X_{141})$".

We draw the channel-based single-queue SBC process algebra Backus-Naur Form tree of A_{141} as shown in Figure 14-2.

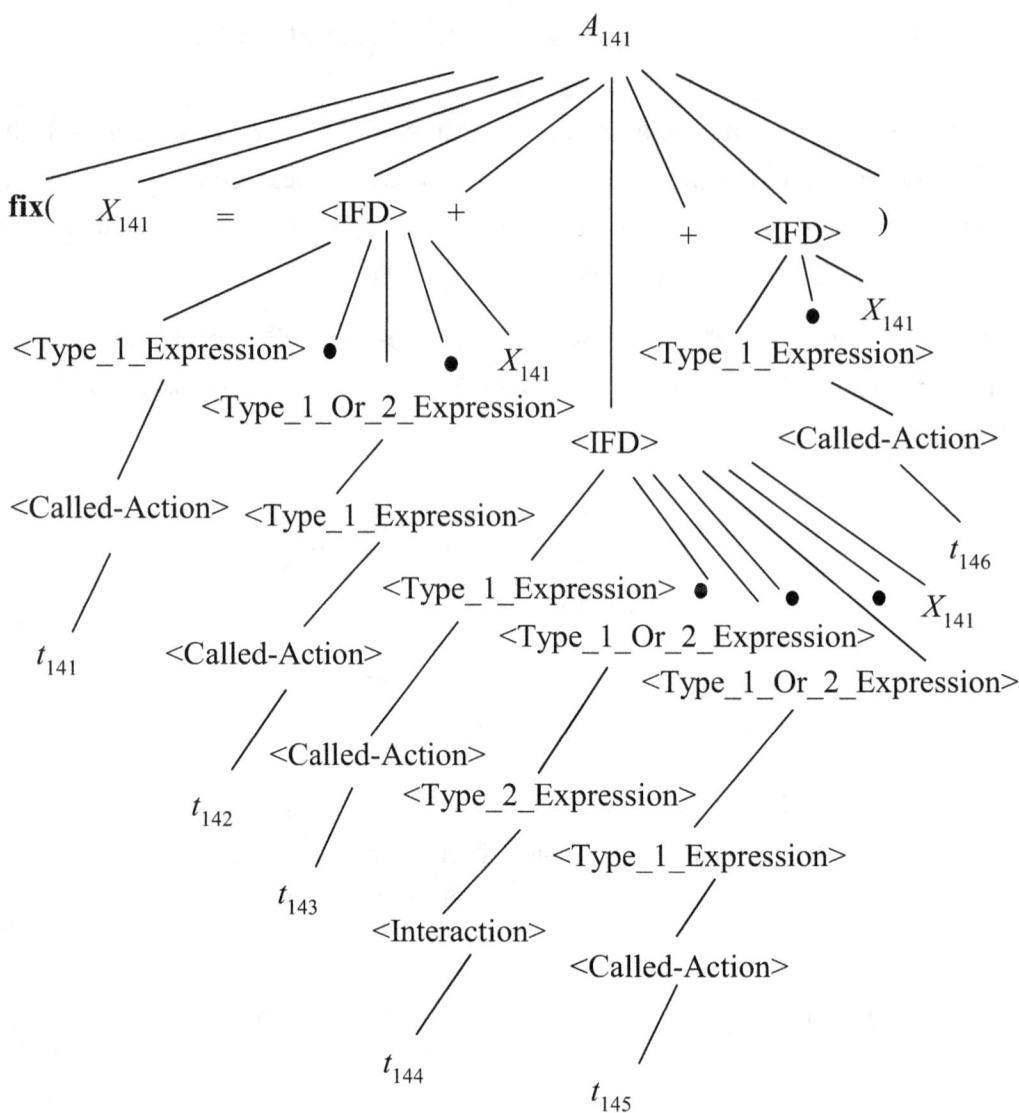

Figure 14-2 Backus-Naur Form Tree of the *Department Store's Car Sweepstakes Sales Promotion System*'s Channel-Based Single-Queue SBC Process

There are three IFDs in the channel-based single-queue SBC process of the *Department Store's Car Sweepstakes Sales Promotion System*. The first IFD is shown in Figure 14-3.

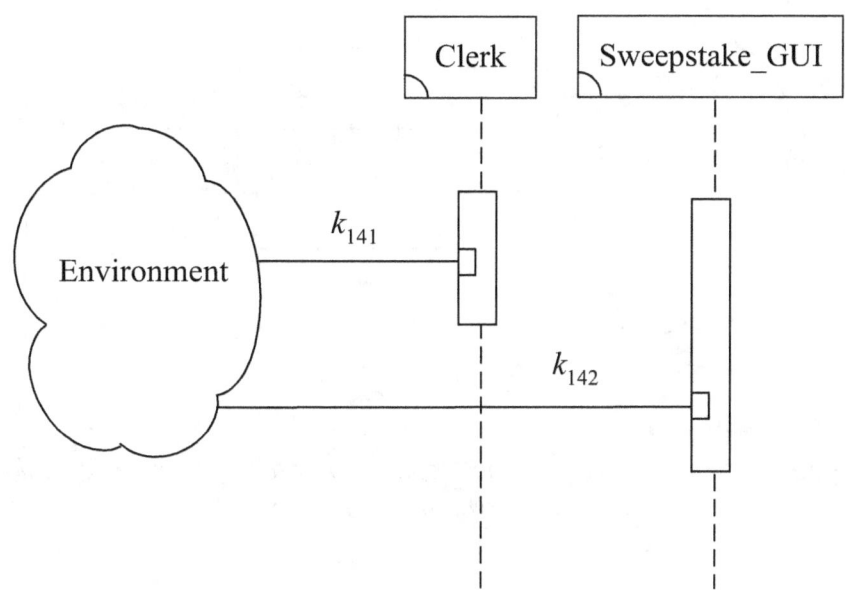

Figure 14-3   First IFD of the *Department Store's Car Sweepstakes Sales Promotion System*

The second IFD of the channel-based single-queue SBC process of the *Department Store's Car Sweepstakes Sales Promotion System* is shown in Figure 14-4.

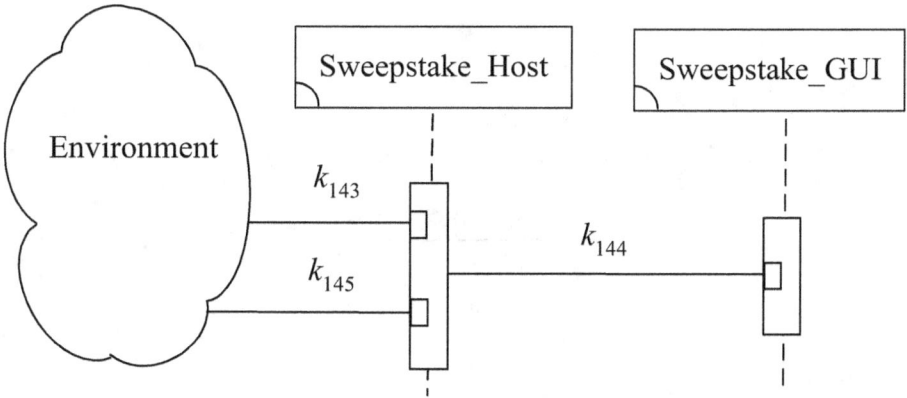

Figure 14-4   Second IFD of the *Department Store's Car Sweepstakes Sales Promotion System*

The third IFD of the channel-based single-queue SBC process of the *Department Store's Car Sweepstakes Sales Promotion System* is shown in Figure 14-5.

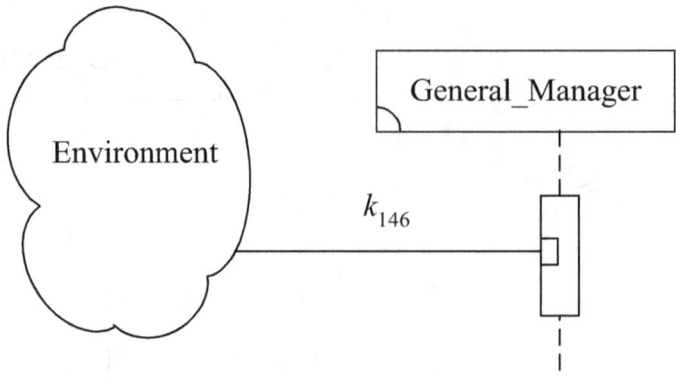

Figure 14-5　Third IFD of the *Department Store's Car Sweepstakes Sales Promotion System*

14-2 Prefixes of the Department Store's Car Sweepstakes Sales Promotion System

The t_{141} (channel-based called action with no conditions and no variable settings) prefix is defined as "<*Clerk*, CALLED, k_{141}>", as shown in Figure 14-6.

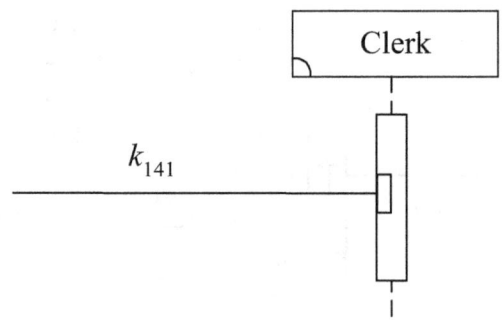

Figure 14-6　Prefix of t_{141}

The t_{142} (channel-based called action with no conditions and no variable settings) prefix is defined as "<$Sweepstake\_GUI$, CALLED, k_{142}>", as shown in Figure 14-7.

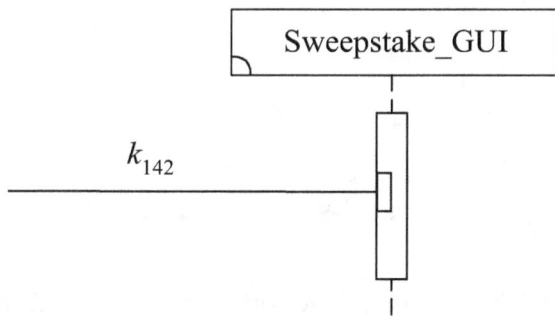

Figure 14-7　Prefix of t_{142}

The t_{143} (channel-based called action with no conditions and no variable settings) prefix is defined as "<$Sweepstake\_Host$, CALLED, k_{143}>", as shown in Figure 14-8.

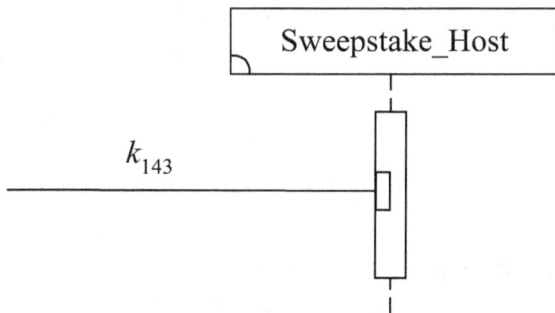

Figure 14-8　Prefix of t_{143}

The t_{144} (channel-based interaction with no conditions and no variable settings) prefix is defined as "<$Sweepstake\_Host$, k_{144}, $Sweepstake\_GUI$>", as shown in Figure 14-9.

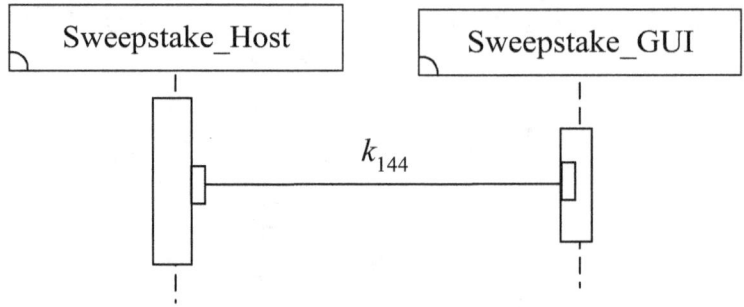

Figure 14-9   Prefix of t_{144}

The t_{145} (channel-based called action with no conditions and no variable settings) prefix is defined as "<*Sweepstake_Host*, CALLED, k_{145}>", as shown in Figure 14-10.

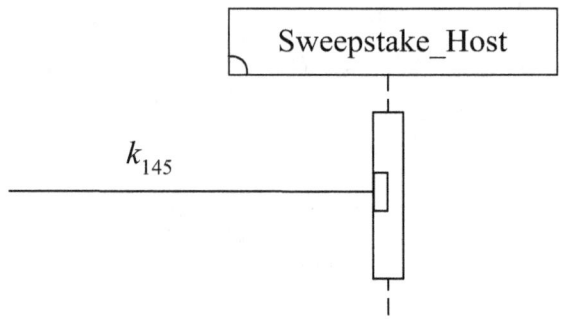

Figure 14-10   Prefix of t_{145}

The t_{146} (channel-based called action with no conditions and no variable settings) prefix is defined as "<*General_Manager*, CALLED, k_{146}>", as shown in Figure 14-11.

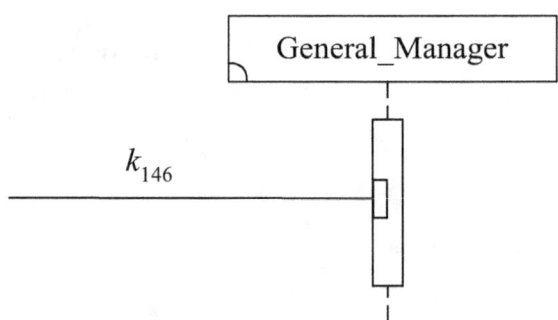

Figure 14-11   Prefix of t_{146}

Figure 14-12 shows all channel formulas of the channel-based single-queue SBC process of the *Department Store's Car Sweepstakes Sales Promotion System.*

Entity name	Channel Formula
k_{141}	Buy_More_Than_Thousand_Dollars(Out Sweepstake_Number)
k_{142}	Register(In Sweepstake_Number, Personal_Data)
k_{143}	Draw_Out_Call
k_{144}	Lucky_Draw(Out Winners_List)
k_{145}	Draw_Out_Return(Out Winners_List)
k_{146}	Award(In Personal_Data; Out Car)

Figure 14-12 Channel Formulas of the *Department Store's Car Sweepstakes Sales Promotion System*'s Process

Figure 14-13 shows the primitive data type specification of the *Sweepstake_Number* and *Car* parameters.

Parameter	Data Type	Instances
Sweepstake_Number	Text	ABC12345
Car	Vehicle	Toyota Camry 2000

Figure 14-13   Primitive Data Type Specification

Figure 14-14 shows the composite data type specification of the *Personal_Data* input parameter occurring in the *Register(In Sweepstake_Number, Personal_Data)* and *Award(In Personal_Data; Out Car)* channel formulas.

Parameter	*Personal_Data*

Data Type	TABLE of
	  ID: Text
	  Customer_Name: Text
	  Phone: Text
	End TABLE ;

Instances	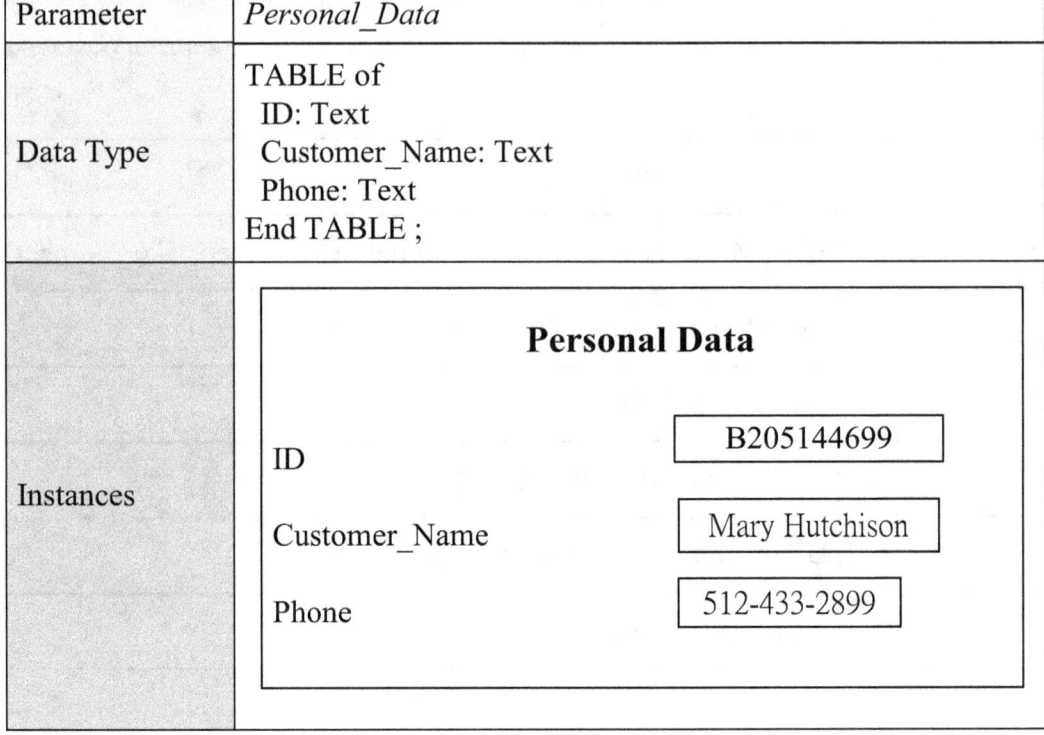

Figure 14-14   Composite Data Type Specification

Figure 14-15 shows the composite data type specification of the *Winners_List)* output parameter occurring in the *Lucky_Draw(Out Winners_List)* and *Draw_Out_Return(Out Winners_List)* channel formulas.

Parameter	*Winners_List*

Data Type	TABLE of
	  ID: Text
	  Customer_Name: Text
	End TABLE ;

Instances	<table><tr><th>ID</th><th>Customer_Name</th></tr><tr><td>B205144699</td><td>Mary Hutchison</td></tr><tr><td>P4305132187</td><td>Alice Bryant</td></tr></table>

Figure 14-15   Composite Data Type Specification

14-3 Process of the Department Store's Car Sweepstakes Sales Promotion System

The following transition graph shows, in Figure 14-16, the semantics of A_{141}'s channel-based single-queue SBC process.

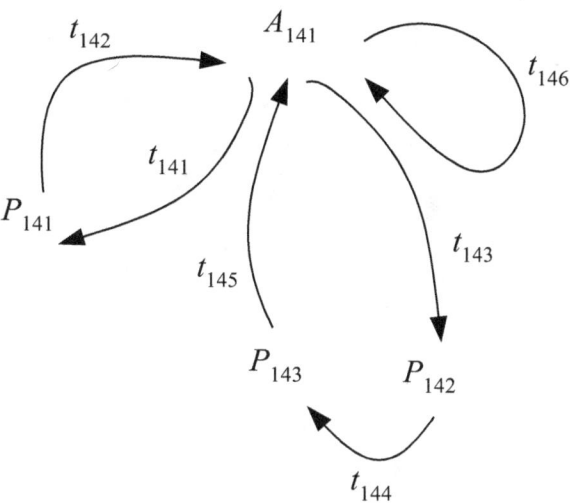

Figure 14-16 Transition graph of the *Department Store's Car Sweepstakes Sales Promotion System*'s Process

In the transition graph of the A_{141}'s channel-based single-queue SBC process, processes A_{141}, P_{141}, P_{142} and P_{143} are defined as in Figure 14-17.

$$A_{141} \stackrel{\text{def}}{=} t_{141} \bullet P_{141} + t_{143} \bullet P_{142} + t_{146} \bullet A_{141}$$
$$P_{141} \stackrel{\text{def}}{=} t_{142} \bullet A_{141}$$
$$P_{142} \stackrel{\text{def}}{=} t_{144} \bullet P_{143}$$
$$P_{143} \stackrel{\text{def}}{=} t_{145} \bullet A_{141}$$

Figure 14-17 Definition of Processes A_{141}, P_{141}, P_{142}, and P_{143}

Chapter 15: Channel-Based Single-Queue SBC Process of the Female Mantis

In this chapter, we use the channel-based single-queue SBC process algebra to model the *Female Mantis* as shown in Figure 15-1.

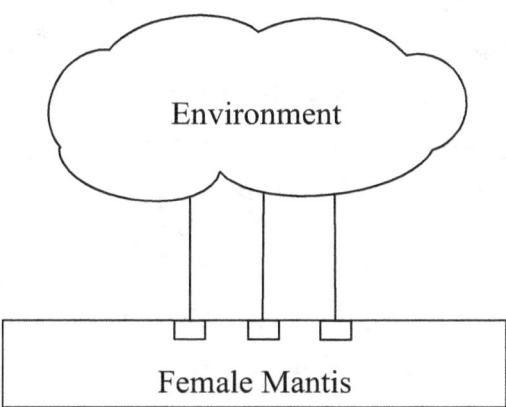

Figure 15-1   Systems Modeling the *Female Mantis*

15-1 BNF Tree of the Female Mantis

The channel-based single-queue SBC process of the *Female Mantis*, A_{201}, is defined as "$\mathbf{fix}(X_{201}=t_{201}\bullet t_{202}\bullet t_{203}\bullet X_{201}+t_{204}\bullet t_{205}\bullet X_{201}+t_{206}\bullet t_{207}\bullet X_{201})$".

We draw the channel-based single-queue SBC process algebra Backus-Naur Form tree of A_{201} as shown in Figure 15-2.

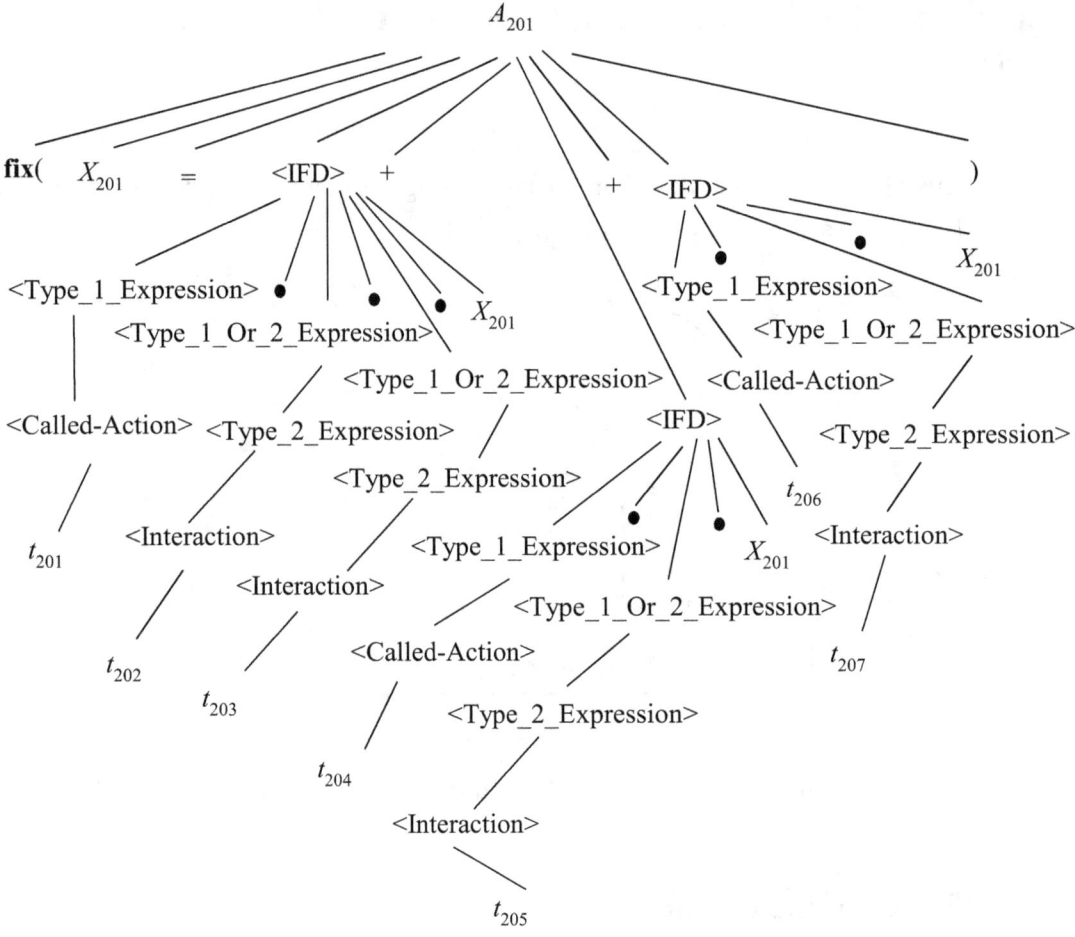

Figure 15-2 Backus-Naur Form Tree of the *Female Mantis'*
Channel-Based Single-Queue SBC Process

There are three IFDs in the channel-based single-queue SBC process of the *Female Mantis*. The first IFD is shown in Figure 15-3.

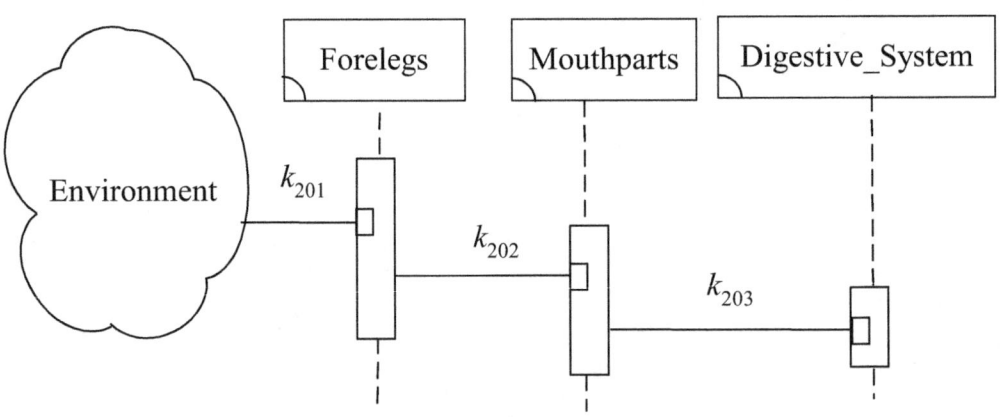

Figure 15-3   First IFD of the *Female Mantis*

The second IFD of the channel-based single-queue SBC process of the *Female Mantis* is shown in Figure 15-4.

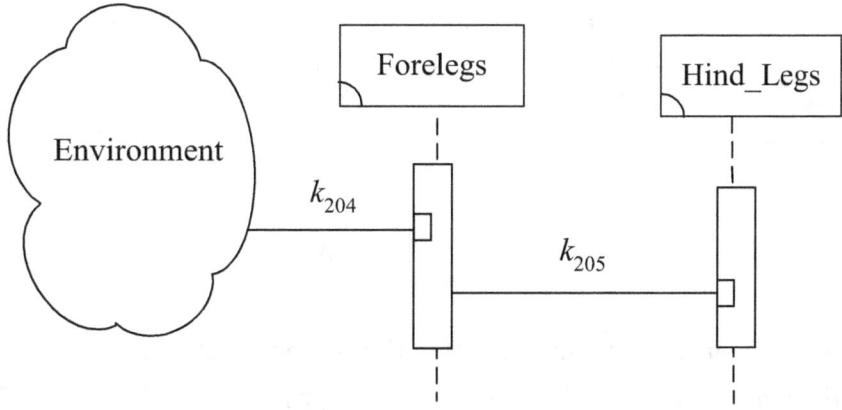

Figure 15-4   Second IFD of the *Female Mantis*

The third IFD of the channel-based single-queue SBC process of the *Female Mantis* is shown in Figure 15-5.

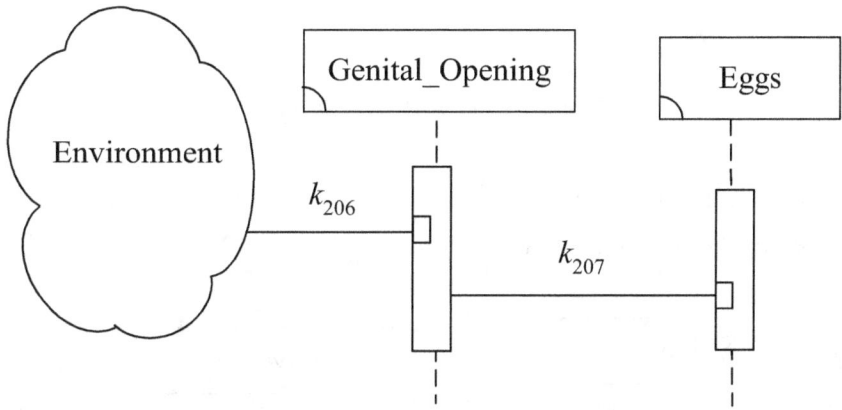

Figure 15-5   Third IFD of the *Female Mantis*

15-2 Prefixes of the Female Mantis

The t_{201} (channel-based called action with no conditions and no variable settings) prefix is defined as "<*Forelegs*, CALLED, k_{201}>", as shown in Figure 15-6.

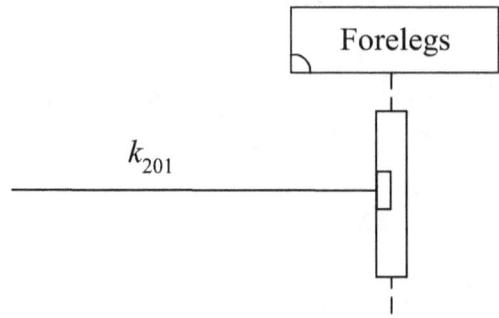

Figure 15-6   Prefix of t_{201}

The t_{202} (channel-based interaction with no conditions and no variable settings) prefix is defined as "<*Forelegs, k_{202}, Mouthparts*>", as shown in Figure 15-7.

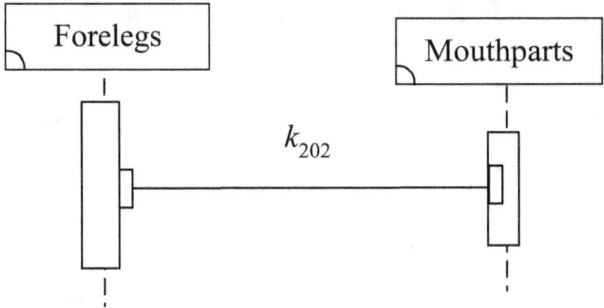

Figure 15-7   Prefix of t_{202}

The t_{203} (channel-based interaction with no conditions and no variable settings) prefix is defined as "<*Mouthparts, k_{203}, Digestive_System*>", as shown in Figure 15-8.

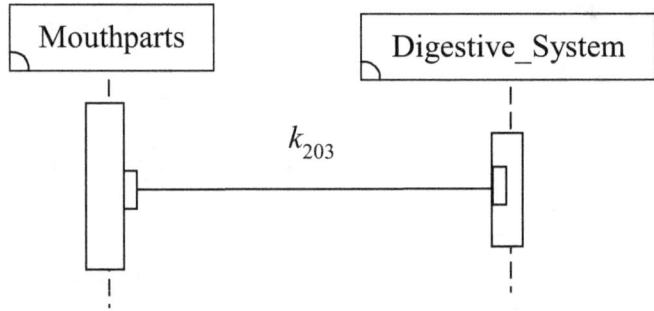

Figure 15-8   Prefix of t_{203}

The t_{204} (channel-based called action with no conditions and no variable settings) prefix is defined as "<*Forelegs*, CALLED, k_{204}>", as shown in Figure 15-9.

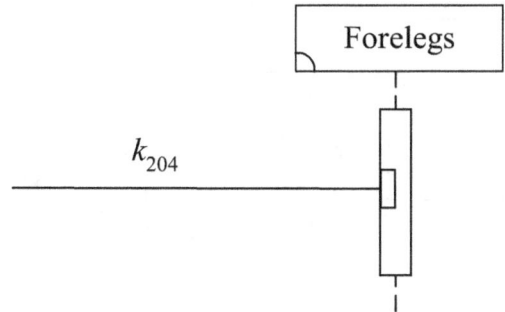

Figure 15-9   Prefix of t_{204}

The t_{205} (channel-based interaction with no conditions and no variable settings) prefix is defined as "<*Forelegs*, k_{205}, *Hind_Legs*>", as shown in Figure 15-10.

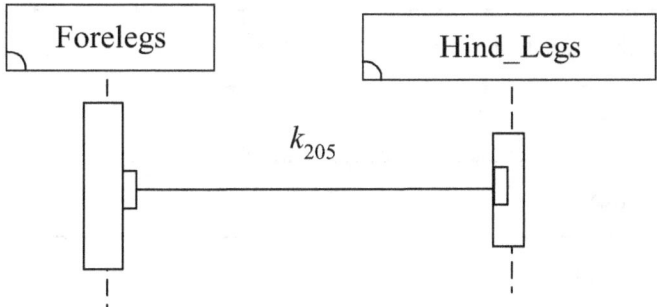

Figure 15-10   Prefix of t_{205}

The t_{206} (channel-based called action with no conditions and no variable settings) prefix is defined as "<*Genital_Opening*, CALLED, k_{206}>", as shown in Figure 15-11.

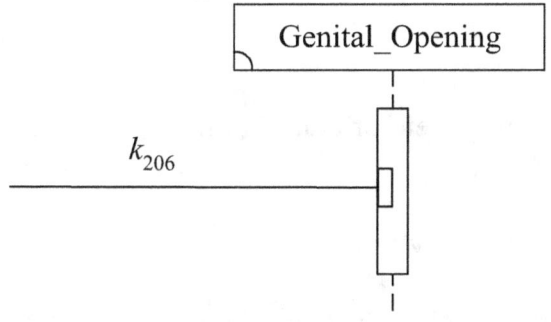

Figure 15-11   Prefix of t_{206}

The t_{207} (channel-based interaction with no conditions and no variable settings) prefix is defined as "<*Genital_Opening*, k_{207}, *Eggs*>", as shown in Figure 15-12.

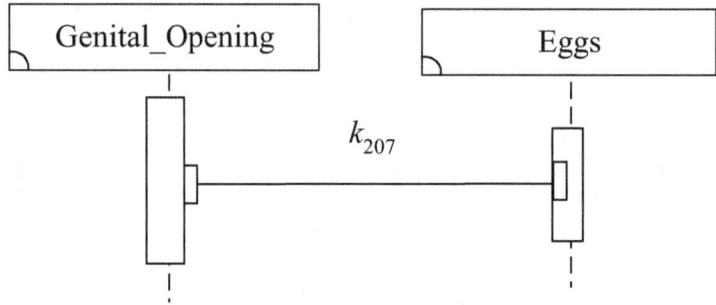

Figure 15-12   Prefix of t_{207}

Figure 15-13 shows all channel formulas of the channel-based single-queue SBC process of the *Female Mantis*.

Entity name	Channel Formula
k_{201}	Grip
k_{202}	Chew
k_{203}	Digest
k_{204}	Moved_forward
k_{205}	Brought_forward
k_{206}	Copulate
k_{207}	Inseminate

Figure 15-13   Channel Formulas of the *Female Mantis*' Process

15-3 Process of the Female Mantis

The following transition graph shows, in Figure 15-14, the semantics of A_{201}'s channel-based single-queue SBC process.

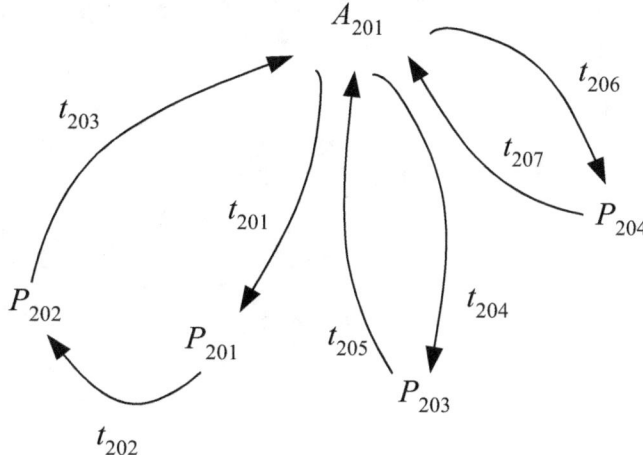

Figure 15-14   Transition graph of the *Female Mantis'* Process

In the transition graph of the A_{201}'s channel-based single-queue SBC process, processes A_{201}, P_{201}, P_{202}, P_{203} and P_{204} are defined as in Figure 15-15.

$$A_{201} \stackrel{def}{=} t_{201} \bullet P_{201} + t_{204} \bullet P_{203} + t_{206} \bullet P_{204}$$

$$P_{201} \stackrel{def}{=} t_{202} \bullet P_{202}$$

$$P_{202} \stackrel{def}{=} t_{203} \bullet A_{201}$$

$$P_{203} \stackrel{def}{=} t_{205} \bullet A_{201}$$

$$P_{204} \stackrel{def}{=} t_{207} \bullet A_{201}$$

Figure 15-15   Definition of Processes A_{201}, P_{201}, P_{202}, P_{203}, and P_{204}

Chapter 16: Channel-Based Single-Queue SBC Process of the Human Body

In this chapter, we use the channel-based single-queue SBC process algebra to model the *Human Body* as shown in Figure 16-1.

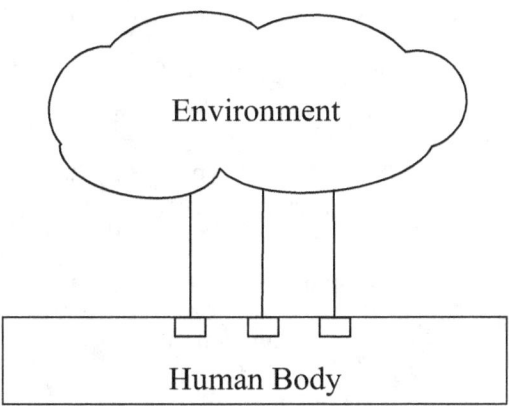

Figure 16-1   Systems Modeling the *Human Body*

16-1 BNF Tree of the Human Body

The channel-based single-queue SBC process of the *Human Body*, A_{251}, is defined as "$\mathbf{fix}(X_{251}=t_{251}\bullet t_{252}\bullet t_{253}\bullet t_{254}\bullet t_{255}\bullet X_{251}+t_{256}\bullet t_{257}\bullet t_{258}\bullet t_{259}\bullet X_{251}+t_{260}\bullet t_{261}\bullet X_{251})$".

We draw the channel-based single-queue SBC process algebra Backus-Naur Form tree of A_{251} as shown in Figure 16-2.

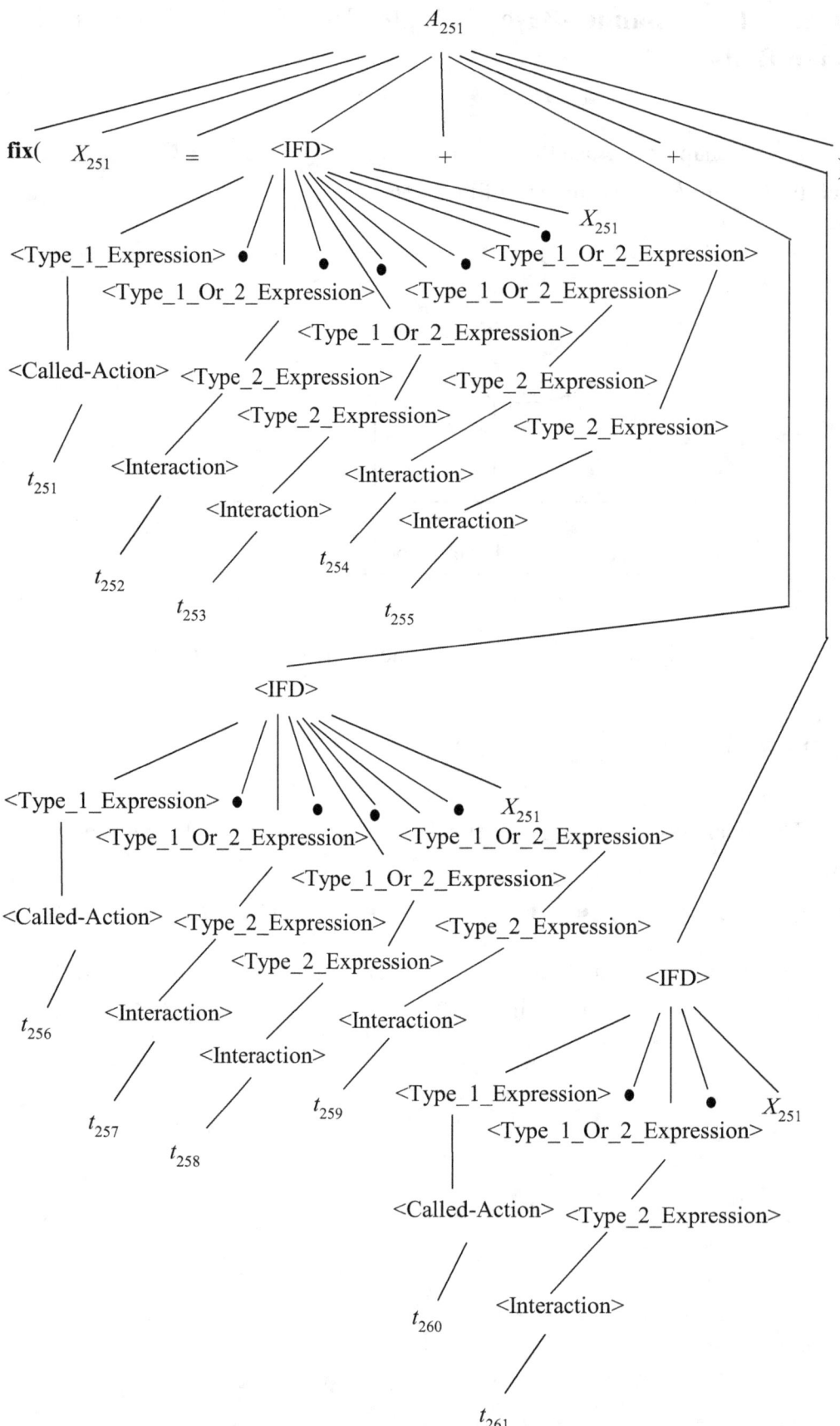

Figure 16-2 Backus-Naur Form Tree of the *Human Body*'s
Channel-Based Single-Queue SBC Process

There are three IFDs in the channel-based single-queue SBC process of the *Human Body*. The first IFD is shown in Figure 16-3.

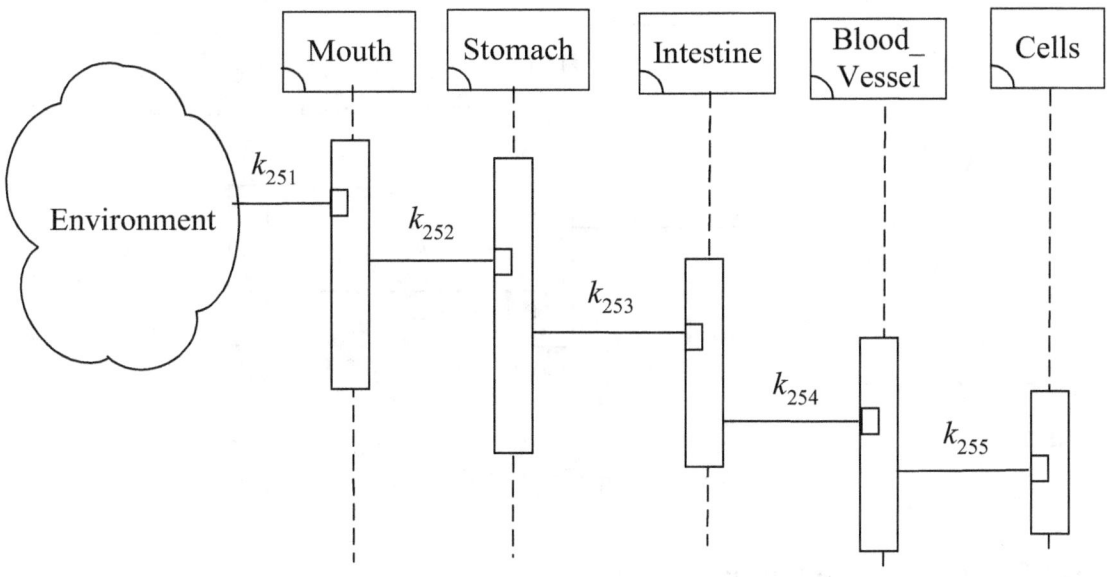

Figure 16-3　　First IFD of the *Human Body*

The second IFD of the channel-based single-queue SBC process of the *Human Body* is shown in Figure 16-4.

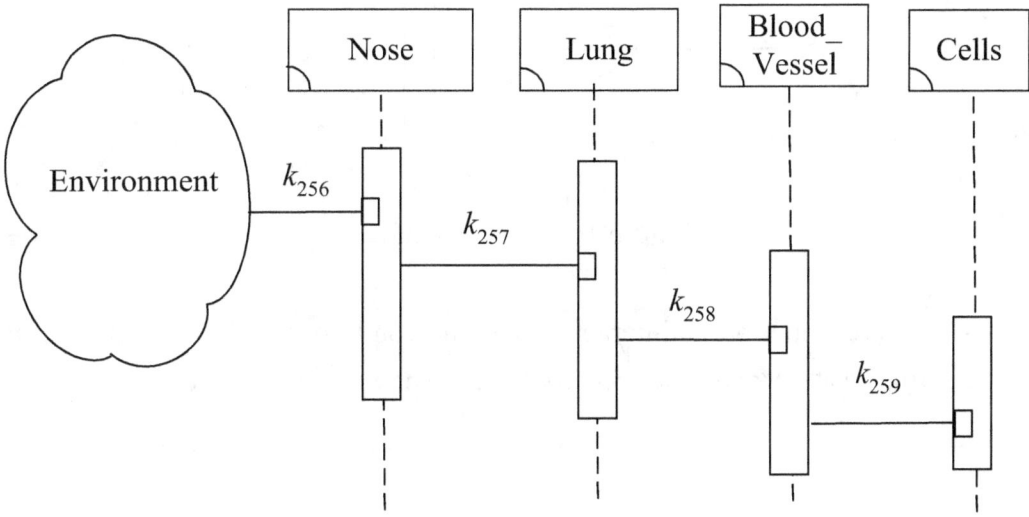

Figure 16-4　　Second IFD of the *Human Body*

The third IFD of the channel-based single-queue SBC process of the *Human Body* is shown in Figure 16-5.

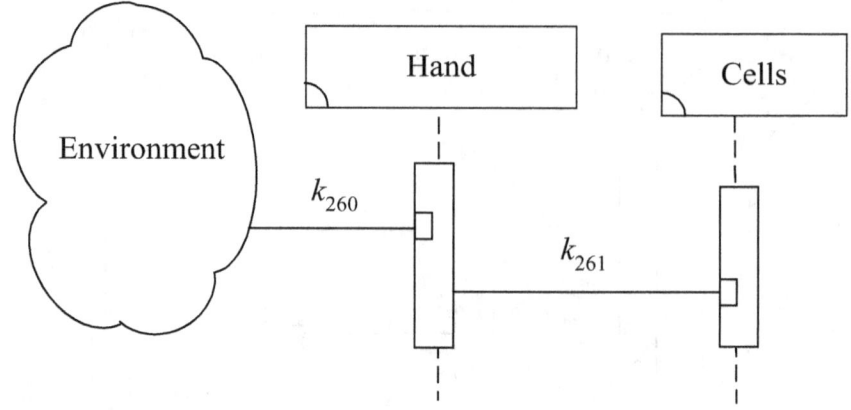

Figure 16-5   Third IFD of the *Human Body*

16-2 Prefixes of the Human Body

The t_{251} (channel-based called action with no conditions and no variable settings) prefix is defined as "<*Mouth*, CALLED, k_{251}>", as shown in Figure 16-6.

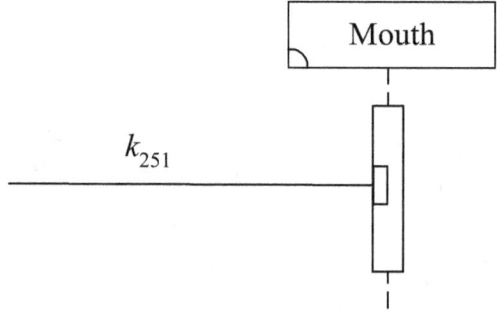

Figure 16-6   Prefix of t_{251}

The t_{252} (channel-based interaction with no conditions and no variable settings) prefix is defined as "<*Mouth*, k_{252}, *Stomach*>", as shown in Figure 16-7.

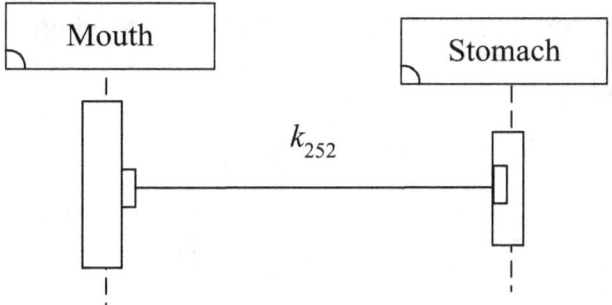

Figure 16-7   Prefix of t_{252}

The t_{253} (channel-based interaction with no conditions and no variable settings) prefix is defined as "<*Stomach*, k_{253}, *Intestine*>", as shown in Figure 16-8.

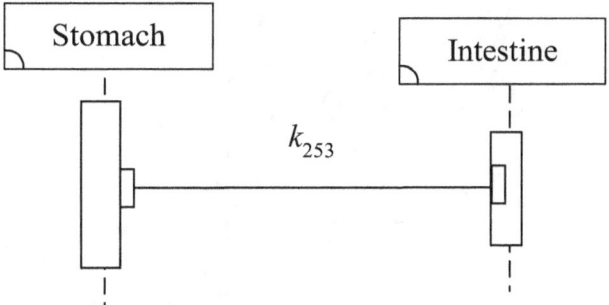

Figure 16-8   Prefix of t_{253}

The t_{254} (channel-based interaction with no conditions and no variable settings) prefix is defined as "<*Intestine*, k_{254}, *Blood_Vessel*>", as shown in Figure 16-9.

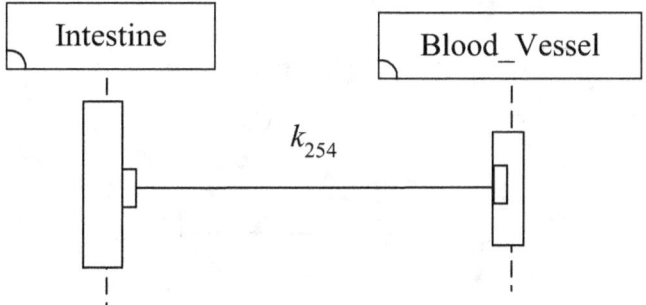

Figure 16-9   Prefix of t_{254}

The t_{255} (channel-based interaction with no conditions and no variable settings) prefix is defined as "<*Blood_Vessel*, k_{255}, *Cells*>", as shown in Figure 16-10.

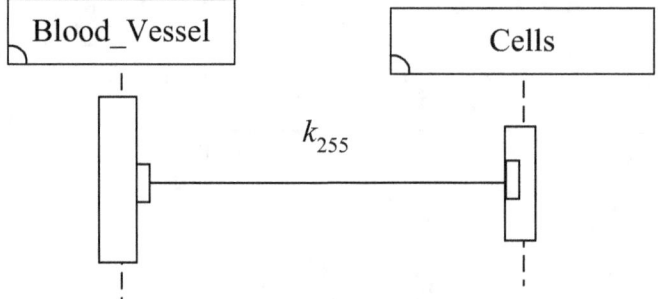

Figure 16-10   Prefix of t_{255}

The t_{256} (channel-based called action with no conditions and no variable settings) prefix is defined as "<*Nose*, CALLED, k_{256}>", as shown in Figure 16-11.

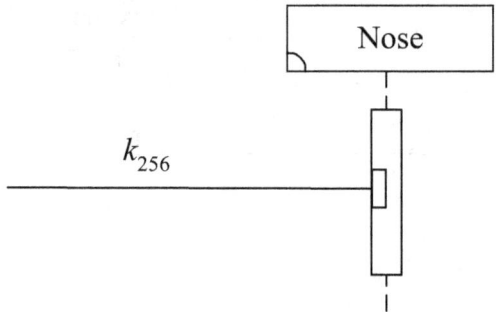

Figure 16-11   Prefix of t_{256}

The t_{257} (channel-based interaction with no conditions and no variable settings) prefix is defined as "<*Nose*, k_{257}, *Lung*>", as shown in Figure 16-12.

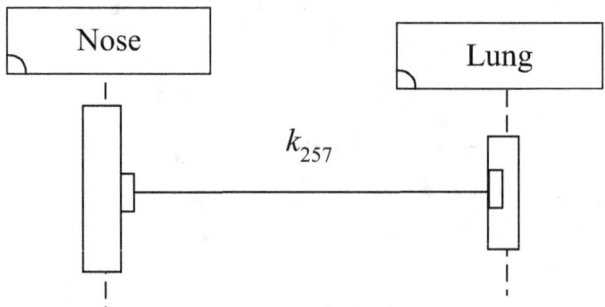

Figure 16-12   Prefix of t_{257}

The t_{258} (channel-based interaction with no conditions and no variable settings) prefix is defined as "<*Lung*, k_{258}, *Blood_Vessel*>", as shown in Figure 16-13.

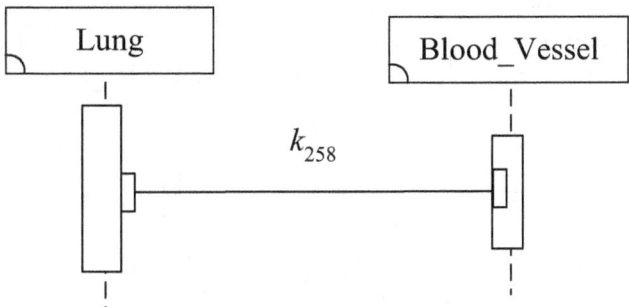

Figure 16-13   Prefix of t_{258}

The t_{259} (channel-based interaction with no conditions and no variable settings) prefix is defined as "<*Blood_Vessel*, k_{259}, *Cells*>", as shown in Figure 16-14.

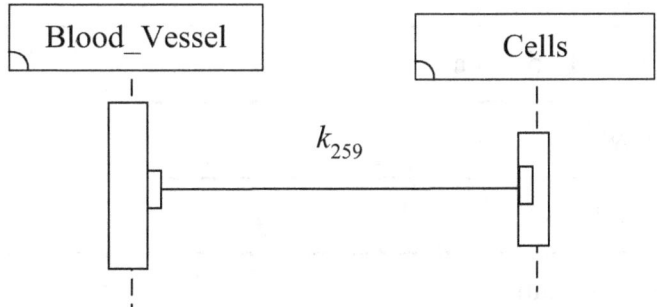

Figure 16-14   Prefix of t_{259}

The t_{260} (channel-based called action with no conditions and no variable settings) prefix is defined as "<*Hand*, CALLED, k_{260}>", as shown in Figure 16-15.

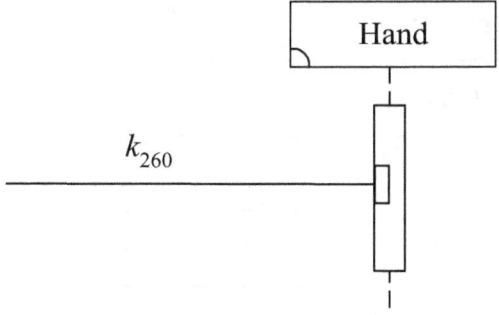

Figure 16-15   Prefix of t_{260}

The t_{261} (channel-based interaction with no conditions and no variable settings) prefix is defined as "<*Hand*, k_{261}, *Cells*>", as shown in Figure 16-16.

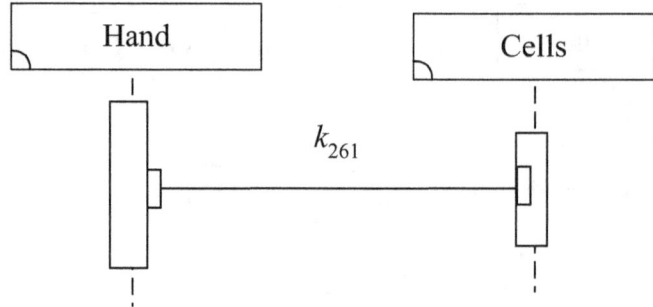

Figure 16-16 Prefix of t_{261}

Figure 16-17 shows all channel formulas of the channel-based single-queue SBC process of the *Human Body*.

Entity name	Channel Formula
k_{251}	Chew
k_{252}	Digest
k_{253}	Absorb_nutrients
k_{254}	Transport_nutrients
k_{255}	Store_nutrients
k_{256}	Breathe
k_{257}	Exchange_gas
k_{258}	Transport_gas
k_{259}	Respire
k_{260}	Punch
k_{261}	Consume_nutrients

Figure 16-17 Channel Formulas of the *Human Body*'s Process

16-3 Process of the Human Body

The following transition graph shows, in Figure 16-18, the semantics of A_{251}'s channel-based single-queue SBC process.

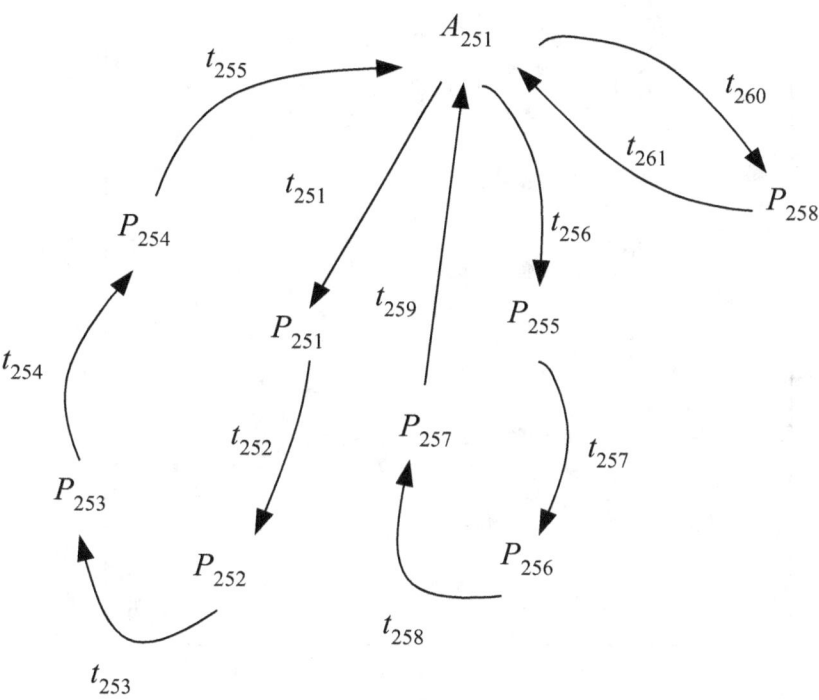

Figure 16-18 Transition graph of the *Human Body*'s Process

In the transition graph of the A_{251}'s channel-based single-queue SBC process, processes A_{251}, P_{251}, P_{252}, P_{253}, P_{254}, P_{255}, P_{256}, P_{257} and P_{258} are defined as in Figure 16-19.

$$A_{251} \stackrel{\text{def}}{=\!=} t_{251} \bullet P_{251} + t_{256} \bullet P_{255} + t_{260} \bullet P_{258}$$

$$P_{251} \stackrel{\text{def}}{=\!=} t_{252} \bullet P_{252}$$

$$P_{252} \stackrel{\text{def}}{=\!=} t_{253} \bullet P_{253}$$

$$P_{253} \stackrel{\text{def}}{=\!=} t_{254} \bullet P_{254}$$

$$P_{254} \stackrel{\text{def}}{=\!=} t_{255} \bullet A_{251}$$

$$P_{255} \stackrel{\text{def}}{=\!=} t_{257} \bullet P_{256}$$

$$P_{256} \stackrel{\text{def}}{=\!=} t_{258} \bullet P_{257}$$

$$P_{257} \stackrel{\text{def}}{=\!=} t_{259} \bullet A_{251}$$

$$P_{258} \stackrel{\text{def}}{=\!=} t_{261} \bullet A_{251}$$

Figure 16-19   Definition of Processes A_{251}, P_{251}, P_{252}, P_{253}, P_{254}, P_{255}, P_{256}, P_{257}, and P_{258}

Chapter 17: Channel-Based Single-Queue SBC Process of the Disaster

In this chapter, we use the channel-based single-queue SBC process algebra to model the *Disaster* as shown in Figure 17-1.

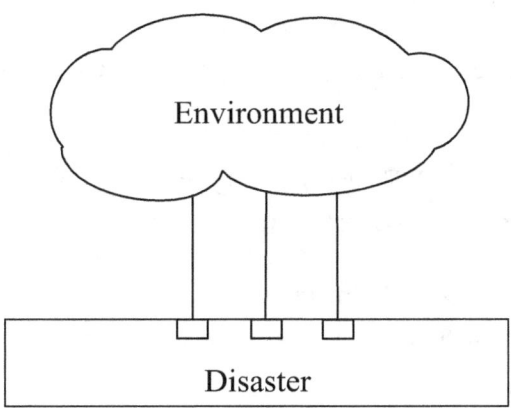

Figure 17-1   Systems Modeling the *Disaster*

17-1 BNF Tree of the Disaster

The channel-based single-queue SBC process of the *Disaster*, A_{301}, is defined as "$\mathbf{fix}(X_{301}=t_{301}\bullet t_{302}\bullet t_{303}\bullet t_{304}\bullet X_{301}+t_{305}\bullet t_{306}\bullet X_{301}+t_{307}\bullet t_{308}\bullet t_{309}\bullet t_{310}\bullet X_{301})$".

We draw the channel-based single-queue SBC process algebra Backus-Naur Form tree of A_{301} as shown in Figure 17-2.

150

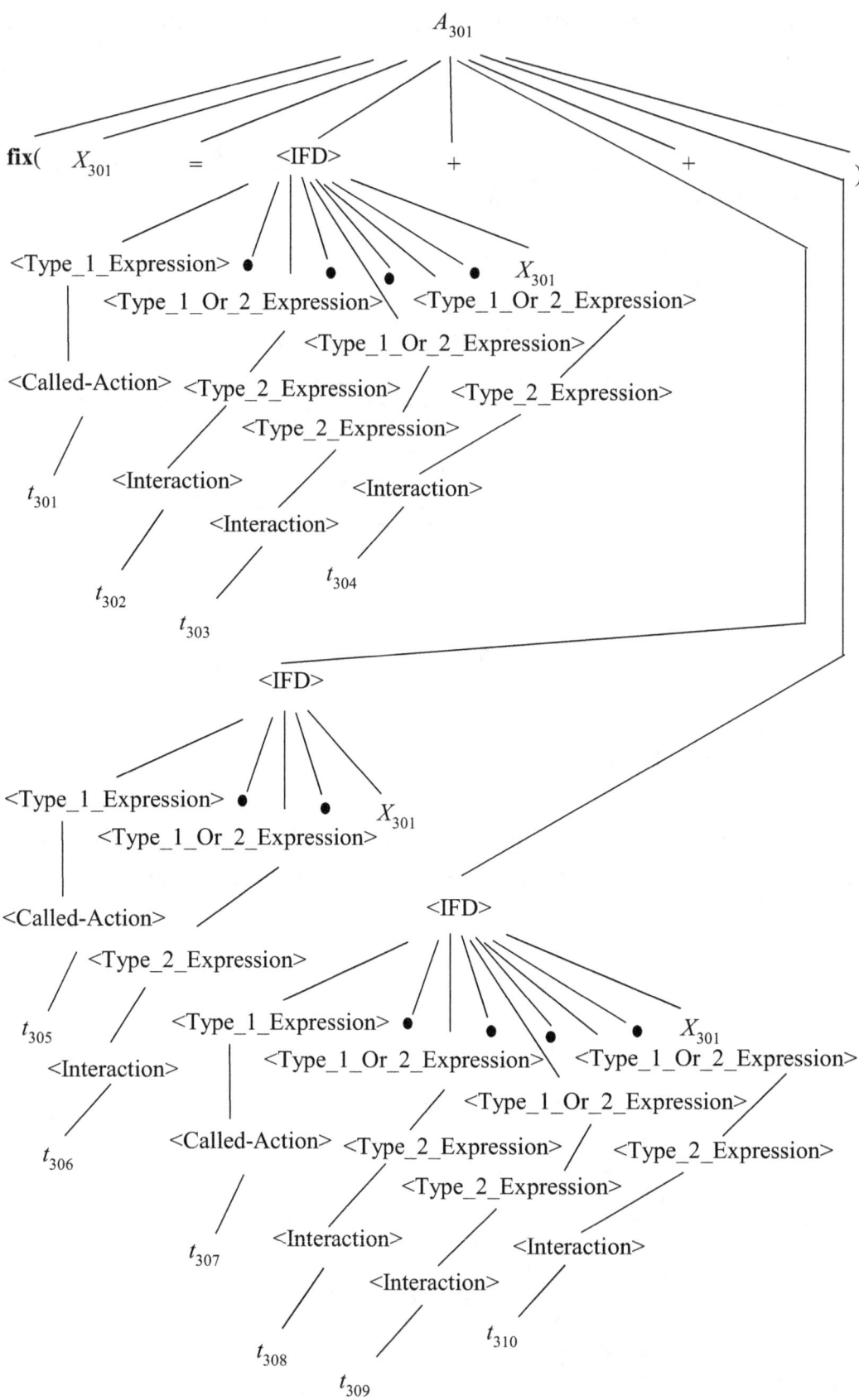

Figure 17-2 Backus-Naur Form Tree of the *Disaster*'s
Channel-Based Single-Queue SBC Process

There are three IFDs in the channel-based single-queue SBC process of the *Disaster*. The first IFD is shown in Figure 17-3.

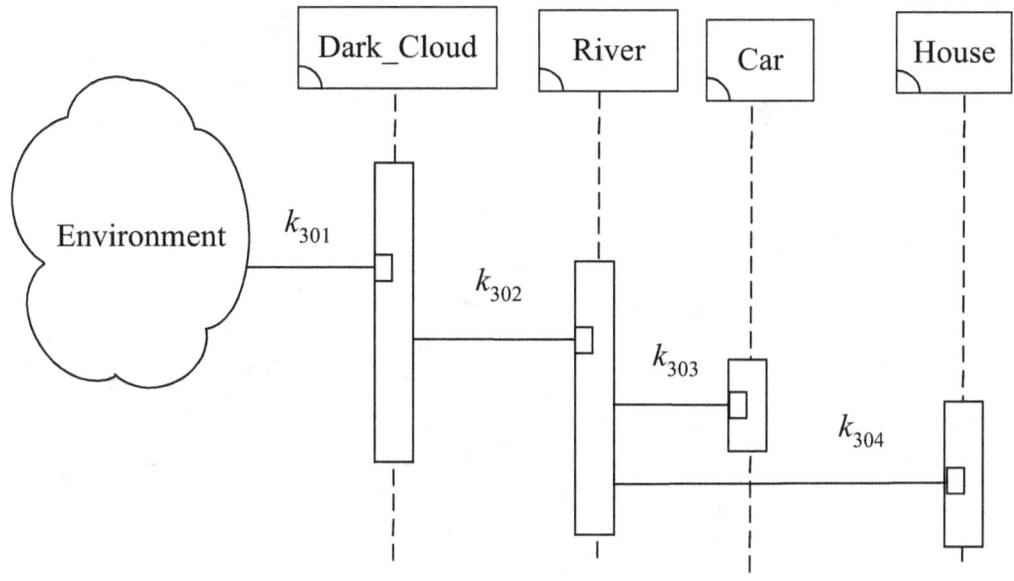

Figure 17-3   First IFD of the *Disaster*

The second IFD of the channel-based single-queue SBC process of the *Disaster* is shown in Figure 17-4.

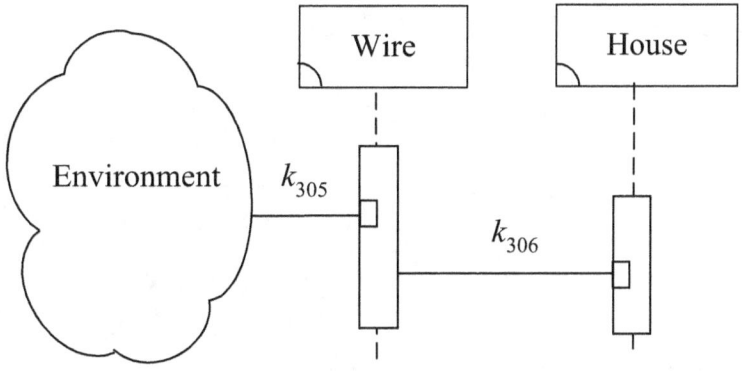

Figure 17-4   Second IFD of the *Disaster*

The third IFD of the channel-based single-queue SBC process of the *Disaster* is shown in Figure 17-5.

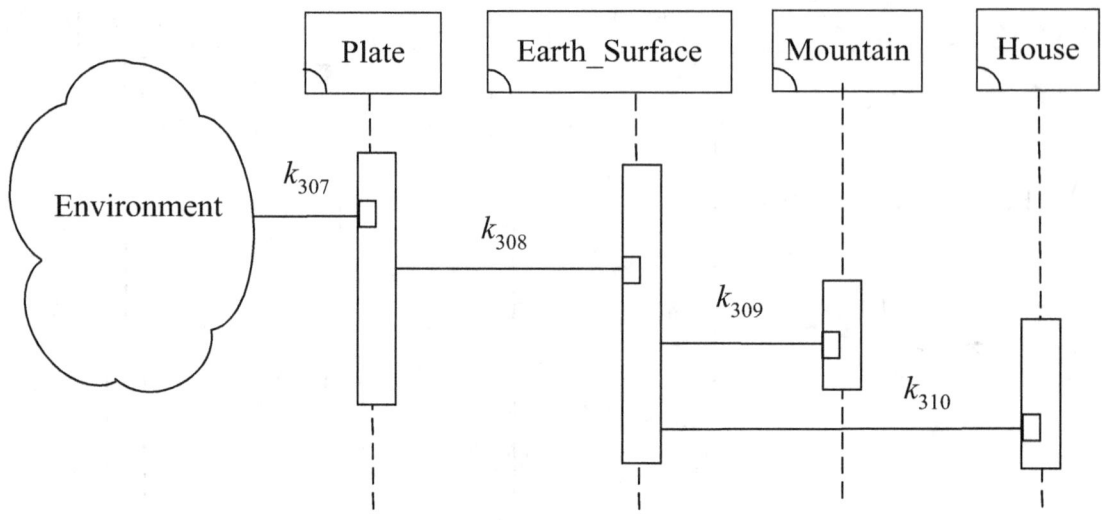

Figure 17-5   Third IFD of the *Disaster*

17-2 Prefixes of the Disaster

The t_{301} (channel-based called action with no conditions and no variable settings) prefix is defined as "<*Dark_Cloud*, CALLED, k_{301}>", as shown in Figure 17-6.

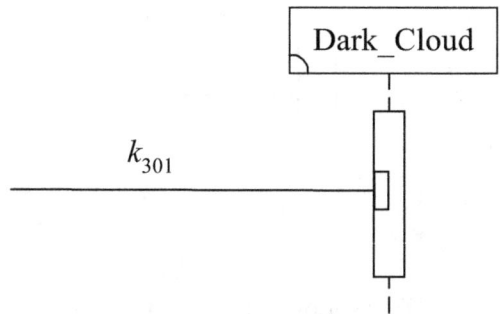

Figure 17-6   Prefix of t_{301}

The t_{302} (channel-based interaction with no conditions and no variable settings) prefix is defined as "<*Dark_Cloud*, k_{302}, *River*>", as shown in Figure 17-7.

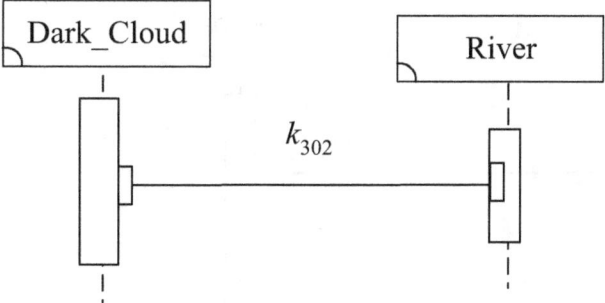

Figure 17-7　Prefix of t_{302}

The t_{303} (channel-based interaction with no conditions and no variable settings) prefix is defined as "<*River*, k_{303}, *Car*>", as shown in Figure 17-8.

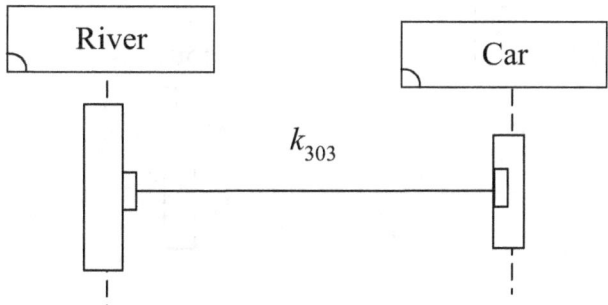

Figure 17-8　Prefix of t_{303}

The t_{304} (channel-based interaction with no conditions and no variable settings) prefix is defined as "<*River*, k_{304}, *House*>", as shown in Figure 17-9.

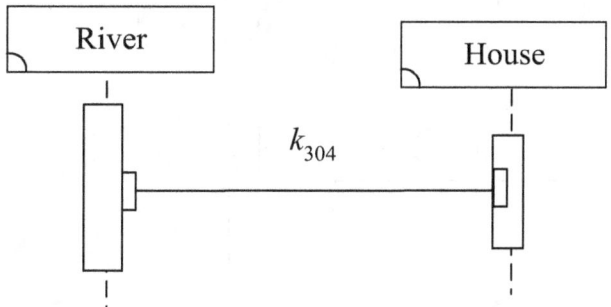

Figure 17-9　Prefix of t_{304}

The t_{305} (channel-based called action with no conditions and no variable settings) prefix is defined as "<*Wire*, CALLED, k_{305}>", as shown in Figure 17-10.

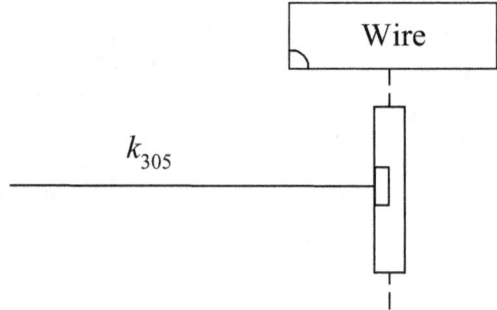

Figure 17-10　Prefix of t_{305}

The t_{306} (channel-based interaction with no conditions and no variable settings) prefix is defined as "<*Wire*, k_{306}, *House*>", as shown in Figure 17-11.

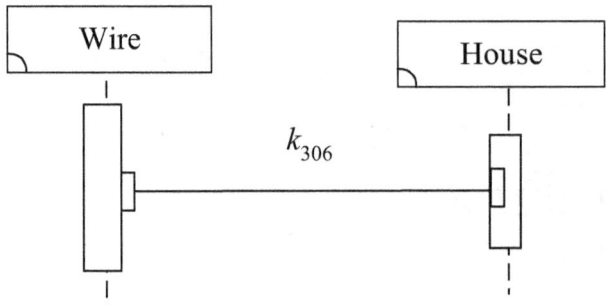

Figure 17-11　Prefix of t_{306}

The t_{307} (channel-based called action with no conditions and no variable settings) prefix is defined as "<*Plate*, CALLED, k_{307}>", as shown in Figure 17-12.

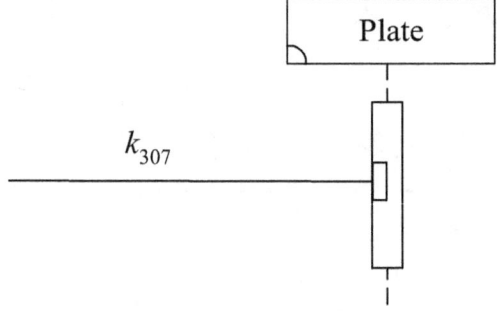

Figure 17-12　Prefix of t_{307}

The t_{308} (channel-based interaction with no conditions and no variable settings) prefix is defined as "<*Plate*, k_{308}, *Earth_Surface*>", as shown in Figure 17-13.

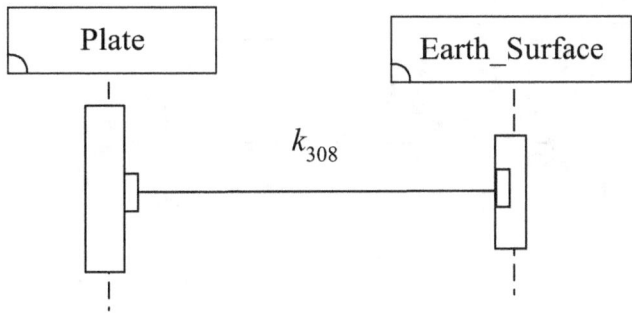

Figure 17-13　Prefix of t_{308}

The t_{309} (channel-based interaction with no conditions and no variable settings) prefix is defined as "<*Earth_Surface*, k_{309}, *Mountain*>", as shown in Figure 17-14.

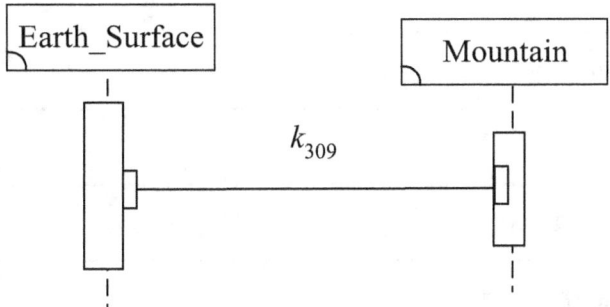

Figure 17-14　Prefix of t_{309}

The t_{310} (channel-based interaction with no conditions and no variable settings) prefix is defined as "<$Earth\_Surface$, k_{310}, $House$>", as shown in Figure 17-15.

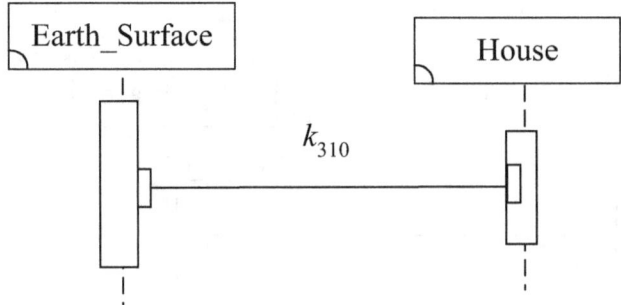

Figure 17-15   Prefix of t_{310}

Figure 17-16 shows all channel formulas of the channel-based single-queue SBC process of the *Disaster*.

Entity name	Channel Formula
k_{301}	Heavy_rain
k_{302}	Overflow
k_{303}	Soak
k_{304}	Wash_away
k_{305}	Bite_through
k_{306}	Burn_down
k_{307}	Move
k_{308}	Ground_rupture
k_{309}	Landslide
k_{310}	Tilt

Figure 17-16   Channel Formulas of the *Disaster*'s Process

17-3 Process of the Disaster

The following transition graph shows, in Figure 17-17, the semantics of A_{301}'s channel-based single-queue SBC process.

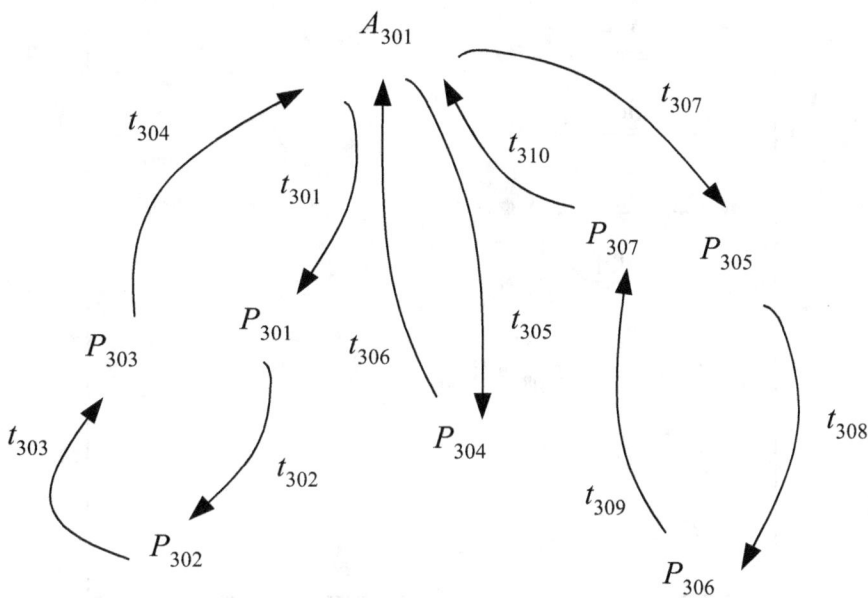

Figure 17-17 Transition graph of the *Disaster*'s Process

In the transition graph of the A_{301}'s channel-based single-queue SBC process, processes $A_{301}, P_{301}, P_{302}, P_{303}, P_{304}, P_{305}, P_{306}$ and P_{307} are defined as in Figure 17-18.

$$A_{301} \stackrel{def}{=} t_{301} \bullet P_{301} + t_{305} \bullet P_{304} + t_{307} \bullet P_{305}$$
$$P_{301} \stackrel{def}{=} t_{302} \bullet P_{302}$$
$$P_{302} \stackrel{def}{=} t_{303} \bullet P_{303}$$
$$P_{303} \stackrel{def}{=} t_{304} \bullet A_{301}$$
$$P_{304} \stackrel{def}{=} t_{306} \bullet A_{301}$$
$$P_{305} \stackrel{def}{=} t_{308} \bullet P_{306}$$
$$P_{306} \stackrel{def}{=} t_{309} \bullet P_{307}$$
$$P_{307} \stackrel{def}{=} t_{310} \bullet A_{301}$$

Figure 17-18   Definition of Processes $A_{301}, P_{301}, P_{302}, P_{303}, P_{304}, P_{305}, P_{306}$, and P_{307}

Chapter 18: Channel-Based Single-Queue SBC Process of the Bicycle

In this chapter, we use the channel-based single-queue SBC process algebra to model the *Bicycle* as shown in Figure 18-1.

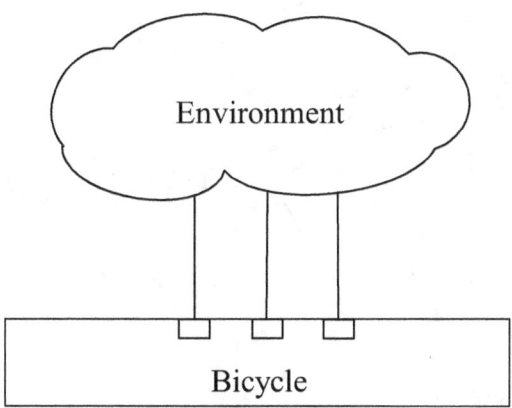

Figure 18-1   Systems Modeling the *Bicycle*

18-1 BNF Tree of the Bicycle

The channel-based single-queue SBC process of the *Bicycle*, A_{401}, is defined as "$\mathbf{fix}(X_{401}=t_{401}\bullet t_{402}\bullet t_{403}\bullet X_{401}+t_{404}\bullet t_{405}\bullet X_{401}+t_{406}\bullet t_{407}\bullet t_{408}\bullet t_{409}\bullet X_{401})$".

We draw the channel-based single-queue SBC process algebra Backus-Naur Form tree of A_{401} as shown in Figure 18-2.

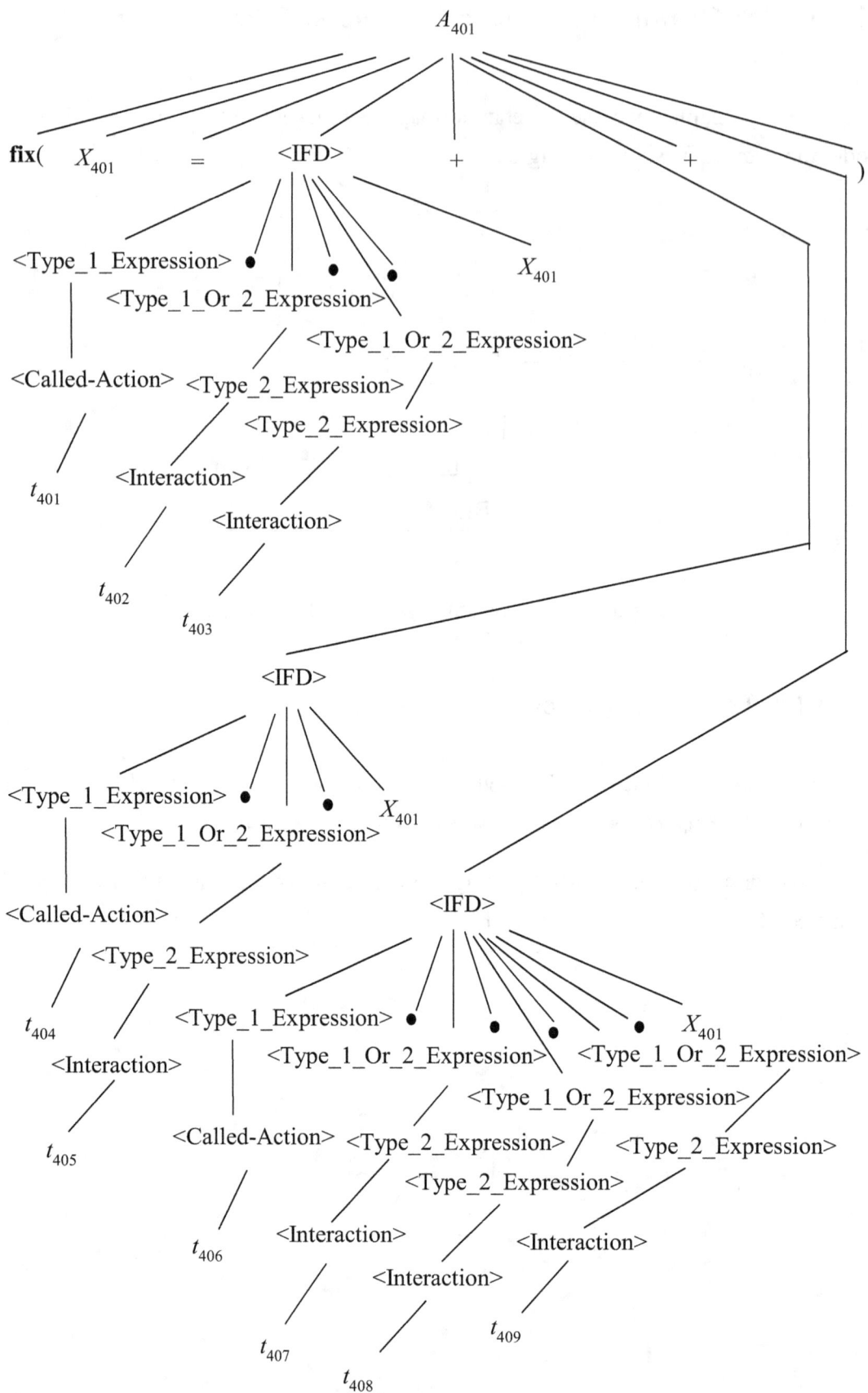

Figure 18-2 Backus-Naur Form Tree of the *Bicycle*'s
Channel-Based Single-Queue SBC Process

There are three IFDs in the channel-based single-queue SBC process of the *Bicycle*. The first IFD is shown in Figure 18-3.

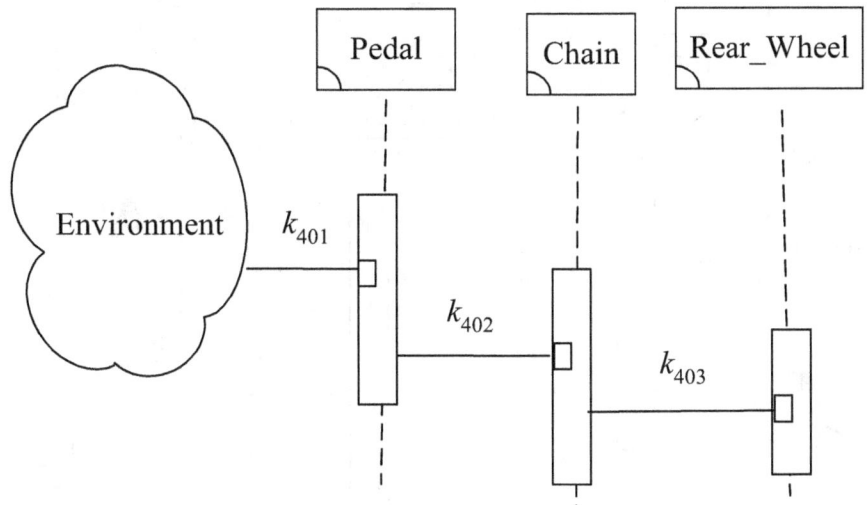

Figure 18-3   First IFD of the *Bicycle*

The second IFD of the channel-based single-queue SBC process of the *Bicycle* is shown in Figure 18-4.

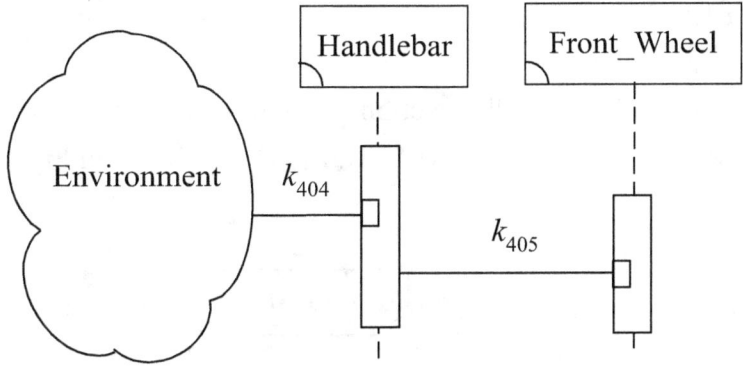

Figure 18-4   Second IFD of the *Bicycle*

The third IFD of the channel-based single-queue SBC process of the *Bicycle* is shown in Figure 18-5.

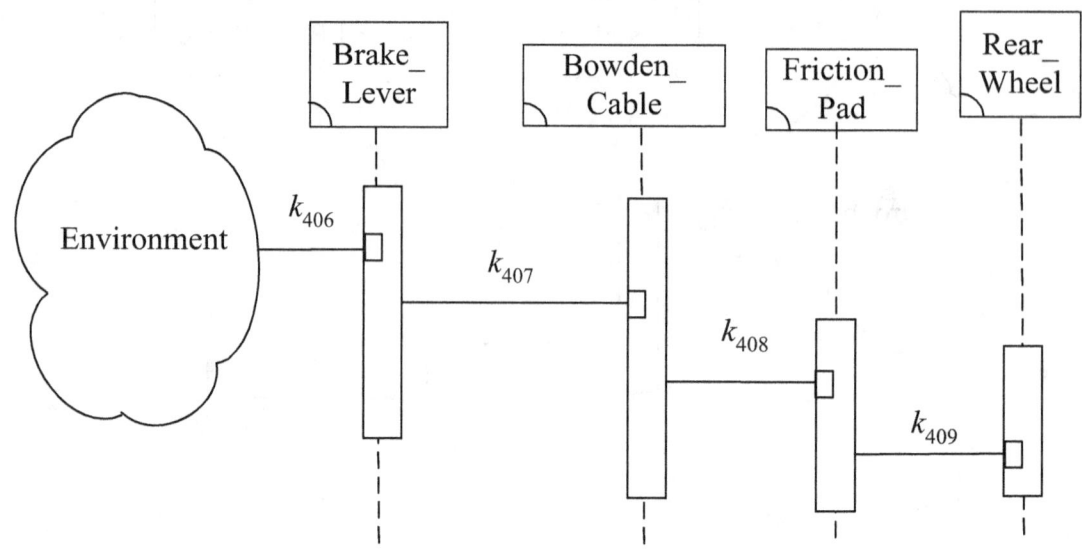

Figure 18-5   Third IFD of the *Bicycle*

18-2 Prefixes of the Bicycle

The t_{401} (channel-based called action with no conditions and no variable settings) prefix is defined as "<*Pedal*, CALLED, k_{401}>", as shown in Figure 18-6.

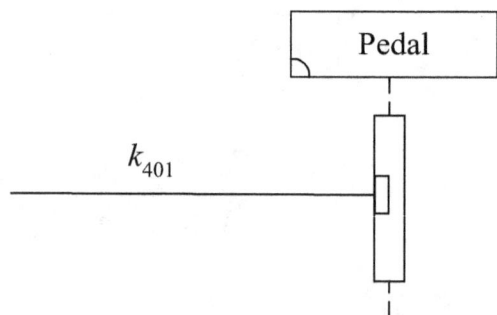

Figure 18-6   Prefix of t_{401}

The t_{402} (channel-based interaction with no conditions and no variable settings) prefix is defined as "<*Pedal*, k_{402}, *Chain*>", as shown in Figure 18-7.

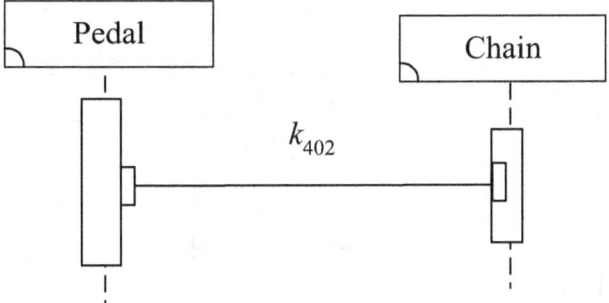

Figure 18-7　Prefix of t_{402}

The t_{403} (channel-based interaction with no conditions and no variable settings) prefix is defined as "<*Chain*, k_{403}, *Rear_Wheel*>", as shown in Figure 18-8.

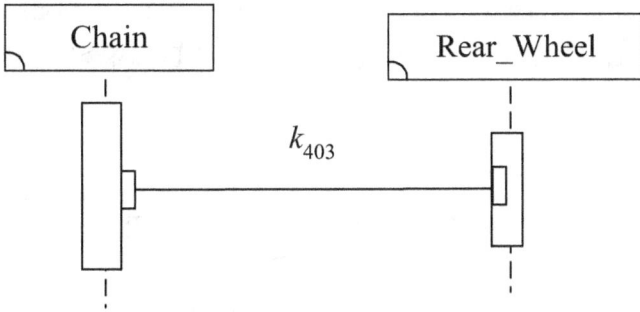

Figure 18-8　Prefix of t_{403}

The t_{404} (channel-based called action with no conditions and no variable settings) prefix is defined as "<*Handlebar*, CALLED, k_{404}>", as shown in Figure 18-9.

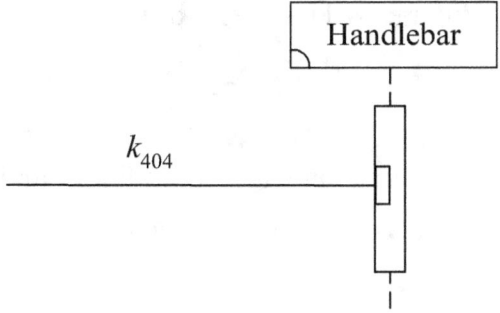

Figure 18-9　Prefix of t_{404}

The t_{405} (channel-based interaction with no conditions and no variable settings) prefix is defined as "<Handlebar, k_{405}, Front_Wheel>", as shown in Figure 18-10.

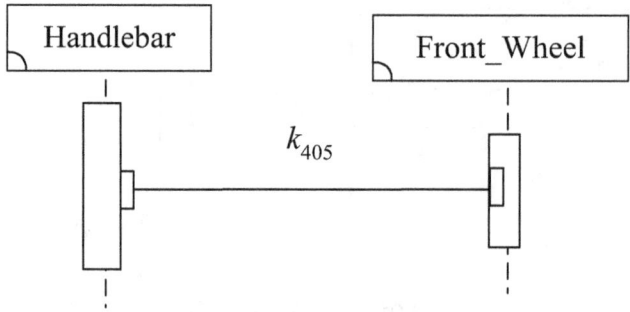

Figure 18-10　Prefix of t_{405}

The t_{406} (channel-based called action with no conditions and no variable settings) prefix is defined as "<Brake_Lever, CALLED, k_{406}>", as shown in Figure 18-11.

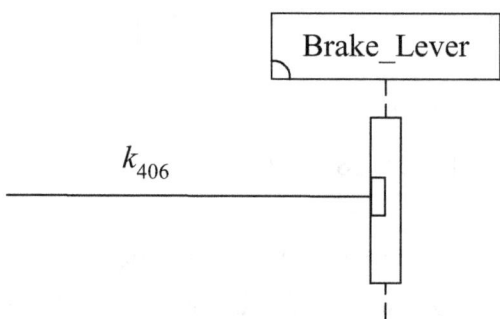

Figure 18-11　Prefix of t_{406}

The t_{407} (channel-based interaction with no conditions and no variable settings) prefix is defined as "<Brake_Lever, k_{407}, Bowden_Cable>", as shown in Figure 18-12.

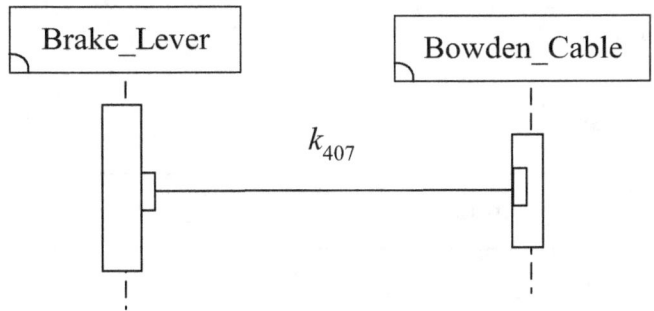

Figure 18-12   Prefix of t_{407}

The t_{408} (channel-based interaction with no conditions and no variable settings) prefix is defined as "<*Bowden_Cable*, k_{408}, *Friction_Pad*>", as shown in Figure 18-13.

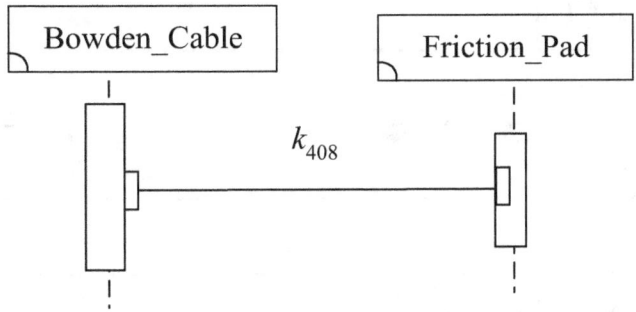

Figure 18-13   Prefix of t_{408}

The t_{409} (channel-based interaction with no conditions and no variable settings) prefix is defined as "<*Friction_Pad*, k_{409}, *Rear_Wheel*>", as shown in Figure 18-14.

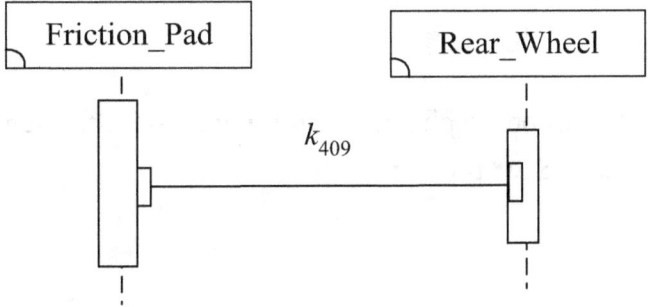

Figure 18-14   Prefix of t_{409}

Figure 18-15 shows all channel formulas of the channel-based single-queue SBC process of the *Bicycle*.

Entity name	Channel Formula
k_{401}	Depress
k_{402}	Transmit_power
k_{403}	Roll
k_{404}	Steer
k_{405}	Turn_left
k_{406}	Force
k_{407}	Pull
k_{408}	Compress
k_{409}	Stop_rolling

Figure 18-15 Channel Formulas of the *Bicycle*'s Process

18-3 Process of the Bicycle

The following transition graph shows, in Figure 18-16, the semantics of A_{401}'s channel-based single-queue SBC process.

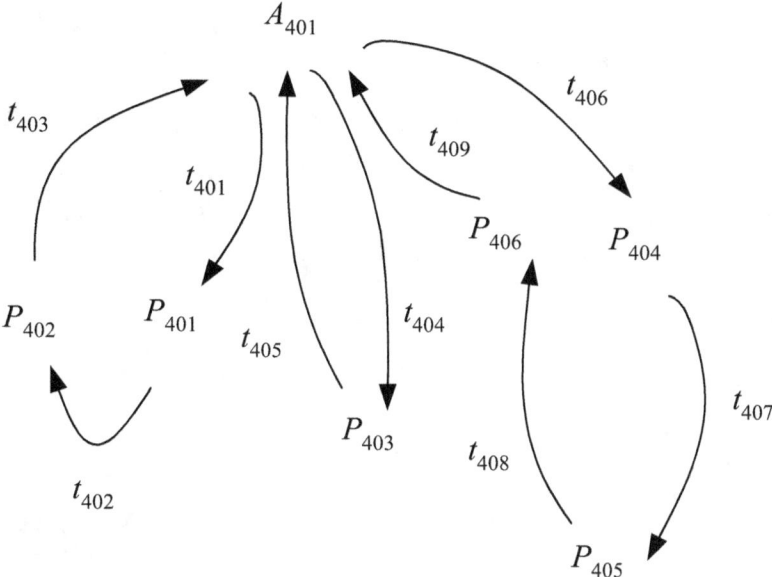

Figure 18-16 Transition graph of the *Bicycle*'s Process

In the transition graph of the A_{401}'s channel-based single-queue SBC process, processes A_{401}, P_{401}, P_{402}, P_{403}, P_{404}, P_{405} and P_{406} are defined as in Figure 18-17.

$$A_{401} \stackrel{\text{def}}{=} t_{401} \bullet P_{401} + t_{404} \bullet P_{403} + t_{406} \bullet P_{404}$$

$$P_{401} \stackrel{\text{def}}{=} t_{402} \bullet P_{402}$$

$$P_{402} \stackrel{\text{def}}{=} t_{403} \bullet A_{401}$$

$$P_{403} \stackrel{\text{def}}{=} t_{405} \bullet A_{401}$$

$$P_{404} \stackrel{\text{def}}{=} t_{407} \bullet P_{405}$$

$$P_{405} \stackrel{\text{def}}{=} t_{408} \bullet P_{406}$$

$$P_{406} \stackrel{\text{def}}{=} t_{409} \bullet A_{401}$$

Figure 18-17 Definition of Processes A_{401}, P_{401}, P_{402}, P_{403}, P_{404}, P_{405}, and P_{406}

Chapter 19: Channel-Based Single-Queue SBC Process of the Restaurant

In this chapter, we use the channel-based single-queue SBC process algebra to model the *Restaurant* as shown in Figure 19-1.

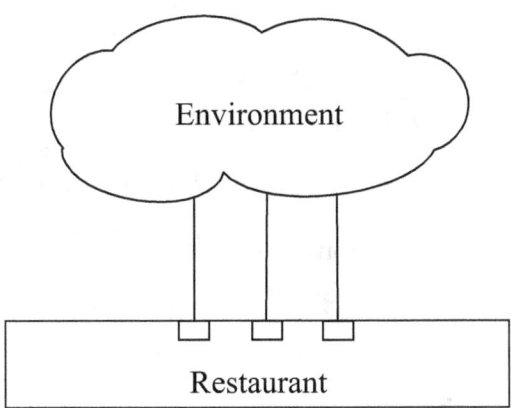

Figure 19-1   Systems Modeling the *Restaurant*

19-1 BNF Tree of the Restaurant

The channel-based single-queue SBC process of the *Restaurant*, A_{501}, is defined as "**fix**$(X_{501}=t_{501}\bullet(t_{502}+t_{503}+t_{504})\bullet(t_{505}+t_{506}+t_{507})\bullet t_{508}\bullet X_{501}+ t_{509}\bullet X_{501})$".

We draw the channel-based single-queue SBC process algebra Backus-Naur Form tree of A_{501} as shown in Figure 19-2.

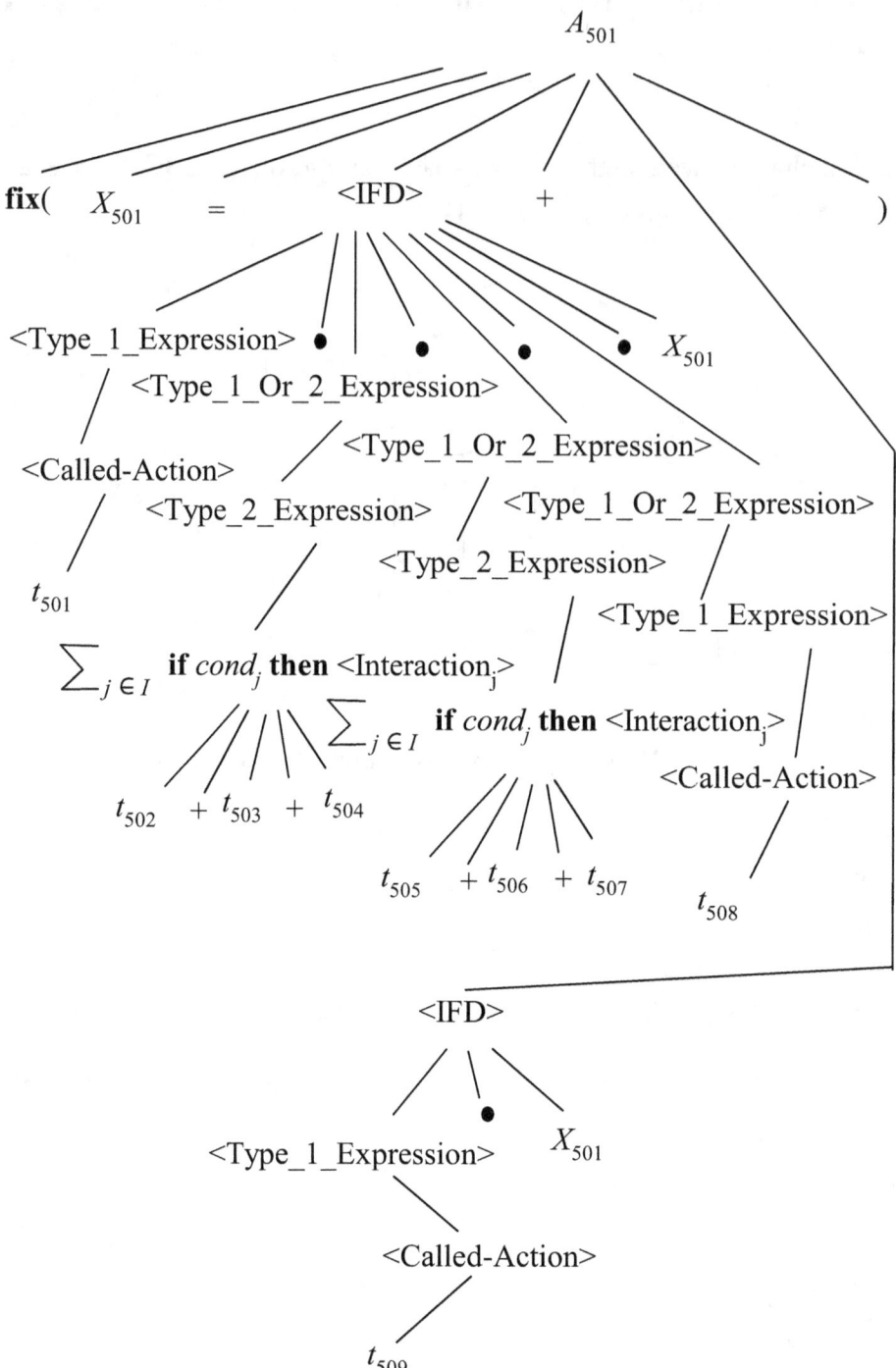

Figure 19-2 Backus-Naur Form Tree of the *Restaurant*'s Channel-Based Single-Queue SBC Process

There are two IFDs in the channel-based single-queue SBC process of the *Restaurant*. The first IFD is shown in Figure 19-3.

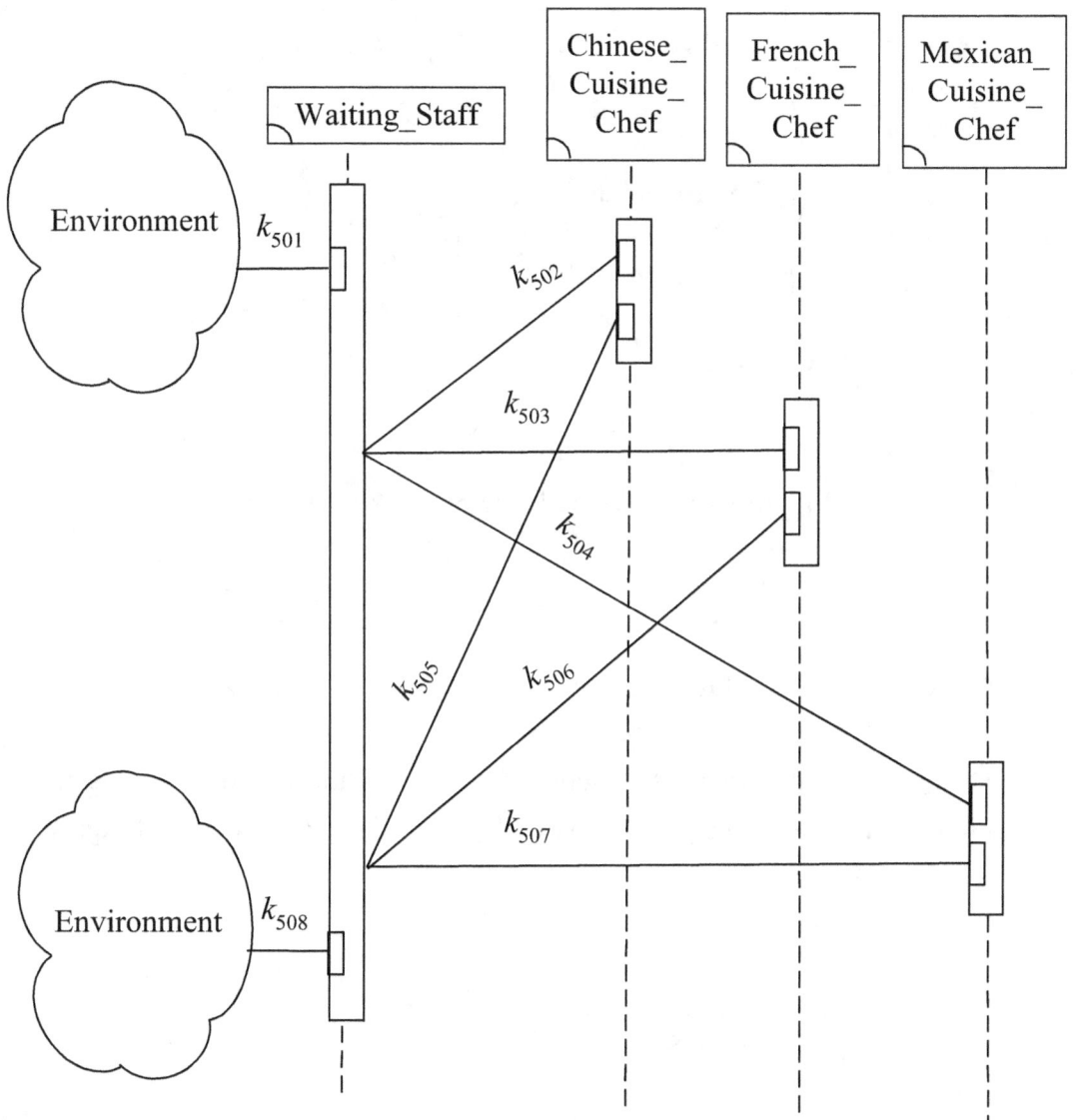

Figure 19-3   First IFD of the *Restaurant*

The second IFD of the channel-based single-queue SBC process of the *Restaurant* is shown in Figure 19-4.

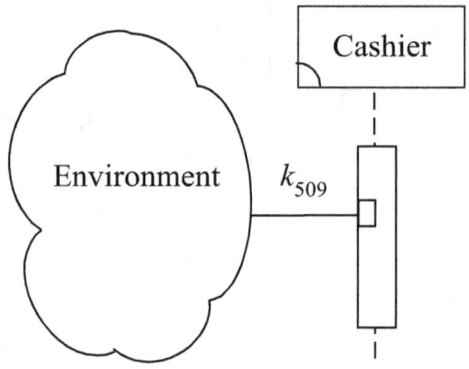

Figure 19-4　　Second IFD of the *Restaurant*

19-2 Prefixes of the Restaurant

The t_{501} (channel-based called action with no conditions and a variable setting) prefix is defined as "<*Waiting_Staff*, CALLED, k_{501}> Order:=Customer_Request", as shown in Figure 19-5.

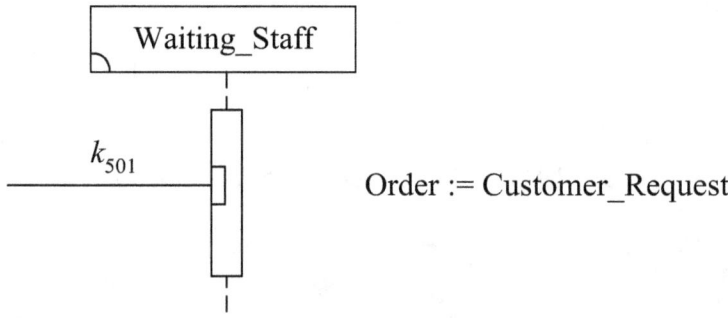

Figure 19-5　　Prefix of t_{501}

The t_{502} (channel-based interaction with a condition and no variable settings) prefix is defined as "**if** Order = "Chinese" **then** <*Waiting_Staff*, k_{502}, *Chinese_Cuisine_Chef*>" as shown in Figure 19-6.

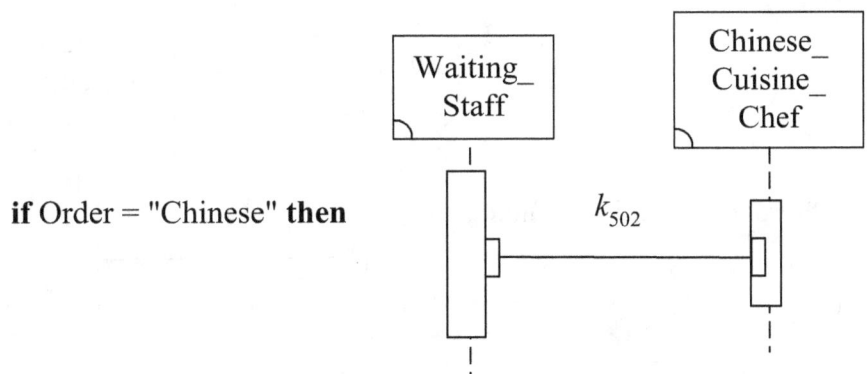

Figure 19-6   Prefix of t_{502}

The t_{503} (channel-based interaction with a condition and no variable settings) prefix is defined as "**if** Order = "French" **then** <*Waiting_Staff*, k_{503}, *French_Cuisine_Chef*>" as shown in Figure 19-7.

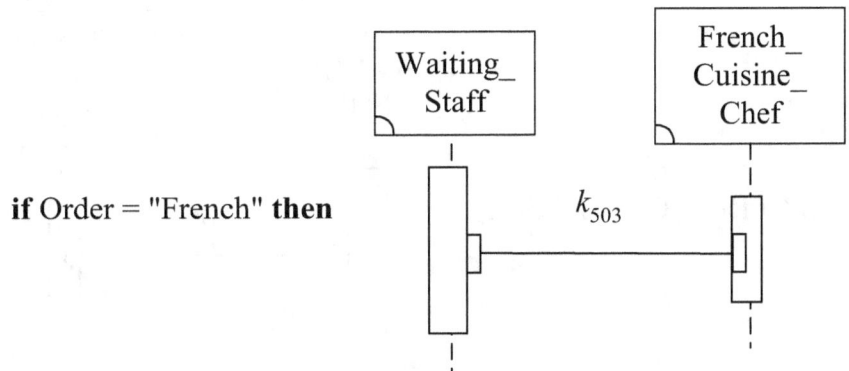

Figure 19-7   Prefix of t_{503}

The t_{504} (channel-based interaction with a condition and no variable settings) prefix is defined as "**if** Order = "Mexican" **then** <*Waiting_Staff*, k_{504}, *Mexican_Cuisine_Chef*>" as shown in Figure 19-8.

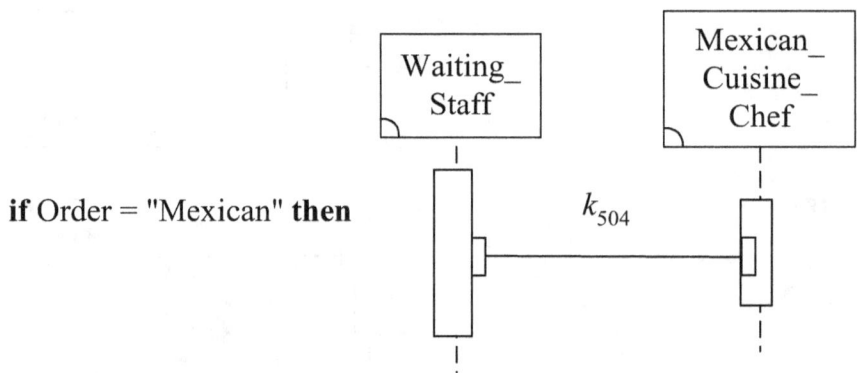

Figure 19-8   Prefix of t_{504}

The t_{505} (channel-based interaction with a condition and no variable settings) prefix is defined as "**if** Order = "Chinese" **then** <*Waiting_Staff*, k_{505}, *Chinese_Cuisine_Chef*>" as shown in Figure 19-9.

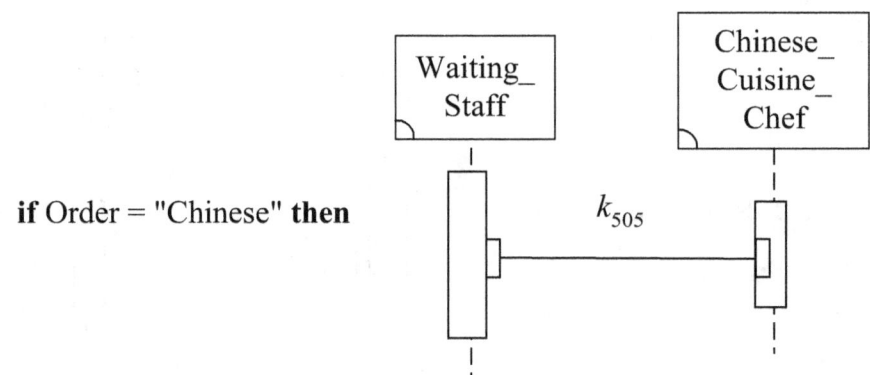

Figure 19-9   Prefix of t_{505}

The t_{506} (channel-based interaction with a condition and no variable settings) prefix is defined as "**if** Order = "French" **then** <*Waiting_Staff*, k_{506}, *French_Cuisine_Chef*>" as shown in Figure 19-10.

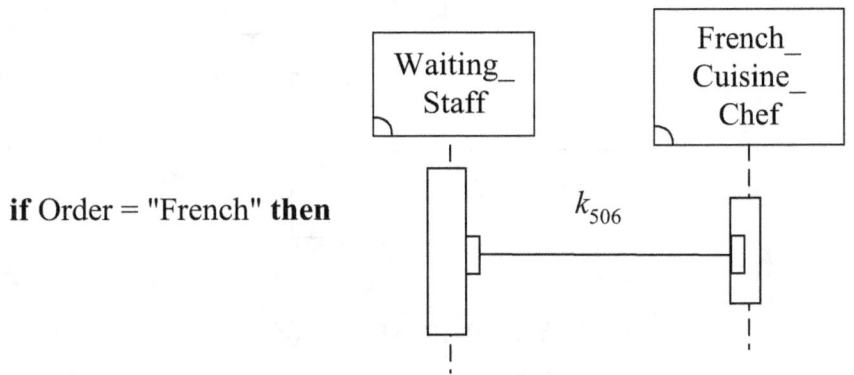

Figure 19-10 Prefix of t_{506}

The t_{507} (channel-based interaction with a condition and no variable settings) prefix is defined as "**if** Order = "Mexican" **then** <*Waiting_Staff*, k_{507}, *Mexican_Cuisine_Chef*>" as shown in Figure 19-11.

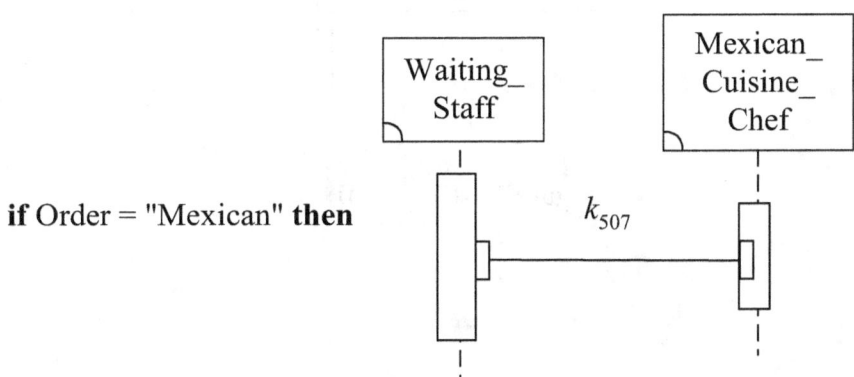

Figure 19-11 Prefix of t_{507}

The t_{508} (channel-based called action with no conditions and no variable settings) prefix is defined as "<*Waiting_Staff*, CALLED, k_{508}>", as shown in Figure 19-12.

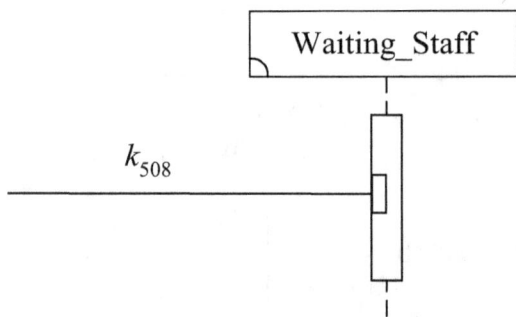

Figure 19-12   Prefix of t_{508}

The t_{509} (channel-based called action with no conditions and no variable settings) prefix is defined as "<*Cashier*, CALLED, k_{509}>", as shown in Figure 19-13.

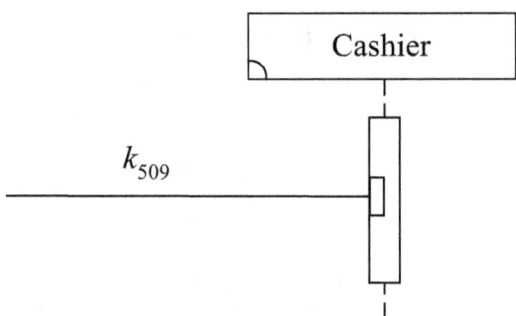

Figure 19-13   Prefix of t_{509}

Figure 19-14 shows all channel formulas of the channel-based single-queue SBC process of the *Restaurant*.

Entity name	Channel Formula
k_{501}	Take_Order_Call(In Customer_Request)
k_{502}	Cook_Chinese_Food_Call
k_{503}	Cook_French_Food_Call
k_{504}	Cook_Mexican_Food_Call
k_{505}	Cook_Chinese_Food_Return(Out Meal_1)
k_{506}	Cook_French_Food_Return(Out Meal_2)
k_{507}	Cook_Mexican_Food_Return(Out Meal_3)
k_{508}	Take_Order_Return(Out Meal)
k_{509}	Pay_Bills

Figure 19-14   Channel Formulas of the *Restaurant*'s Process

Figure 19-15 shows the primitive data type specification of the *Order* variable, the *Customer_Request* input parameter, and the *Meal, Meal_1, Meal_2, Meal_3* output parameters.

Parameter	Data Type	Instances
Order	Enumerated	Chinese, French, Mexican
Customer_Request	Enumerated	Chinese, French, Mexican
Meal	Enumerated food	Chinese food, French food, Mexican food
Meal_1	Food	Chinese food
Meal_2	Food	French food
Meal_3	Food	Mexican food

Figure 19-15   Primitive Data Type Specification

19-3 Process of the Restaurant

The following transition graph shows, in Figure 19-16, the semantics of A_{501}'s channel-based single-queue SBC process.

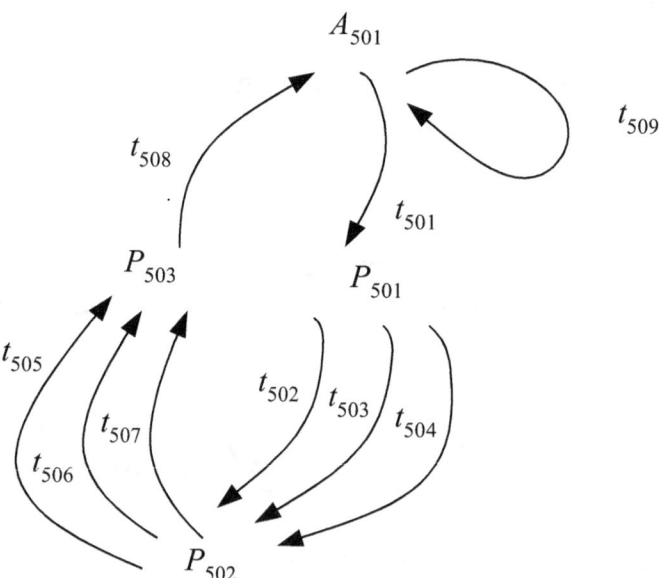

Figure 19-16   Transition graph of the *Restaurant*'s Process

In the transition graph of the A_{501}'s channel-based single-queue SBC process, processes A_{501}, P_{501}, P_{502} and P_{503} are defined as in Figure 19-17.

$$A_{501} \stackrel{\text{def}}{=\joinrel=} t_{501} \bullet P_{501} + t_{509} \bullet A_{501}$$

$$P_{501} \stackrel{\text{def}}{=\joinrel=} t_{502} \bullet P_{502} + t_{503} \bullet P_{502} + t_{504} \bullet P_{502}$$

$$P_{502} \stackrel{\text{def}}{=\joinrel=} t_{505} \bullet P_{503} + t_{506} \bullet P_{503} + t_{507} \bullet P_{503}$$

$$P_{503} \stackrel{\text{def}}{=\joinrel=} t_{508} \bullet A_{501}$$

Figure 19-17   Definition of Processes A_{501}, P_{501}, P_{502}, and P_{503}

Chapter 20: Channel-Based Single-Queue SBC Process of the Car

In this chapter, we use the channel-based single-queue SBC process algebra to model the *Car* as shown in Figure 20-1.

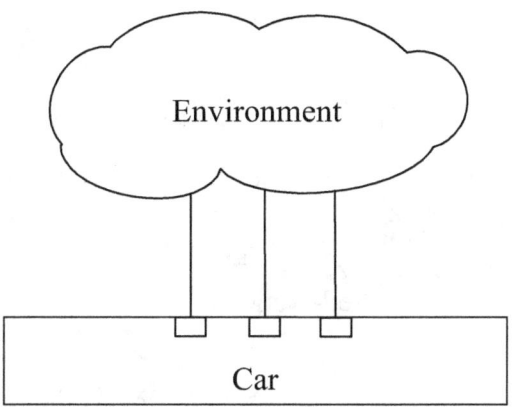

Figure 20-1   Systems Modeling the *Car*

20-1 BNF Tree of the Car

The channel-based single-queue SBC process of the *Car*, A_{601}, is defined as "$\mathbf{fix}(X_{601}=t_{601}\bullet X_{601}+ t_{602}\bullet X_{601}+t_{603}\bullet t_{604}\bullet(t_{605}+t_{606})\bullet X_{601})$".

We draw the channel-based single-queue SBC process algebra Backus-Naur Form tree of A_{601} as shown in Figure 20-2.

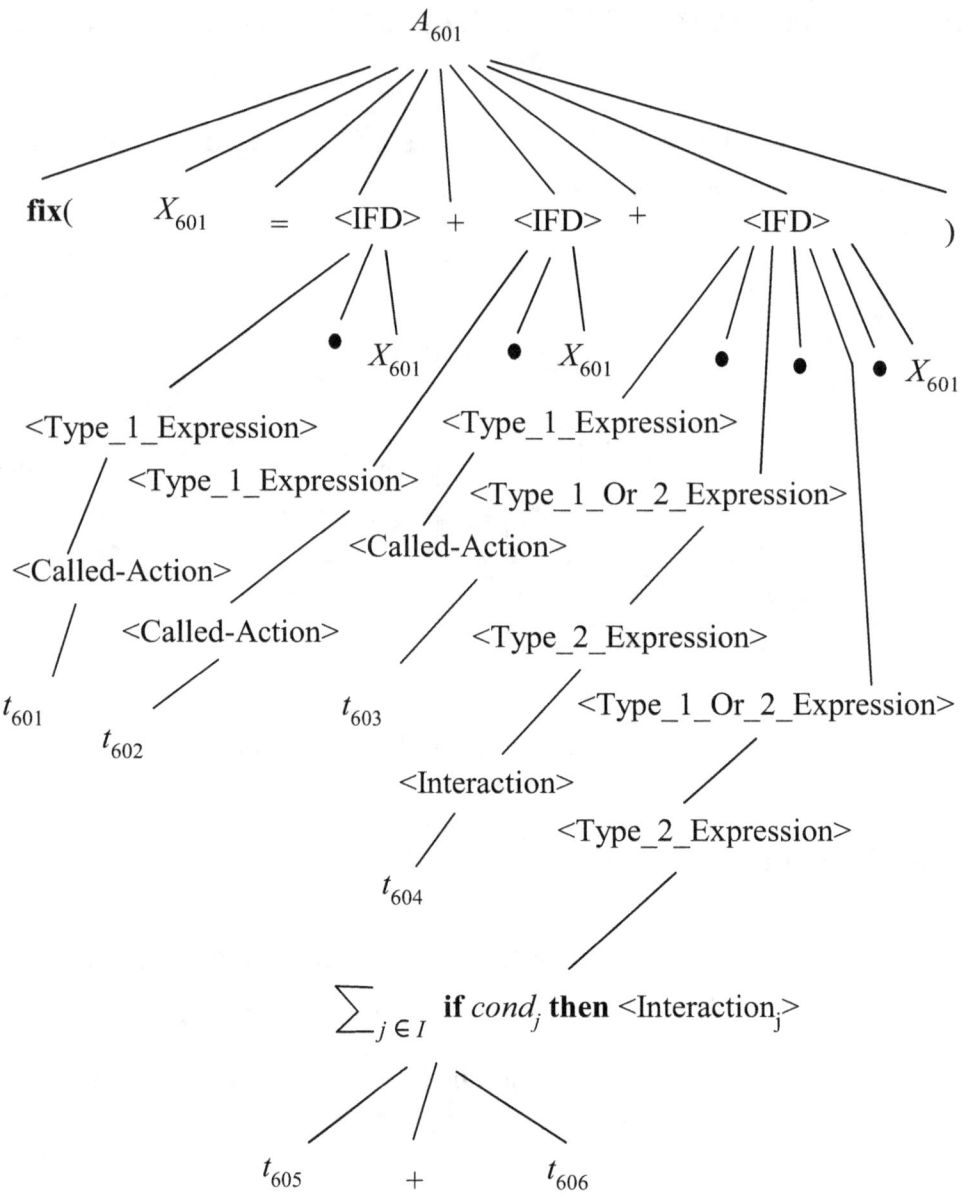

Figure 20-2 Backus-Naur Form Tree of the *Car*'s
Channel-Based Single-Queue SBC Process

There are three IFDs in the channel-based single-queue SBC process of the *Car*. The first IFD is shown in Figure 20-3.

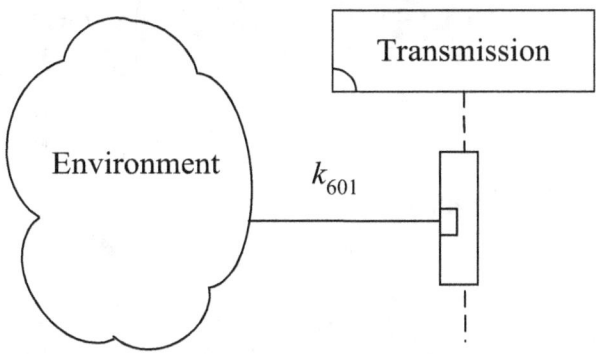

Figure 20-3　First IFD of the *Car*

The second IFD of the channel-based single-queue SBC process of the *Car* is shown in Figure 20-4.

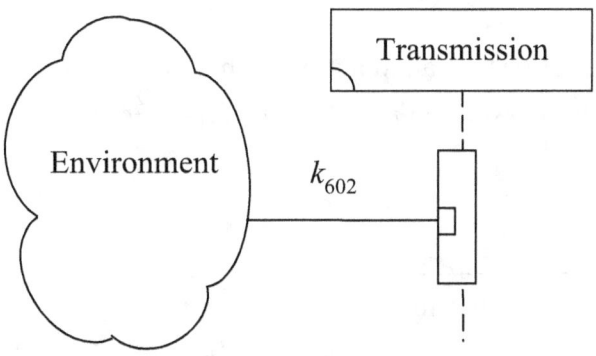

Figure 20-4　Second IFD of the *Car*

The third IFD of the channel-based single-queue SBC process of the *Car* is shown in Figure 20-5.

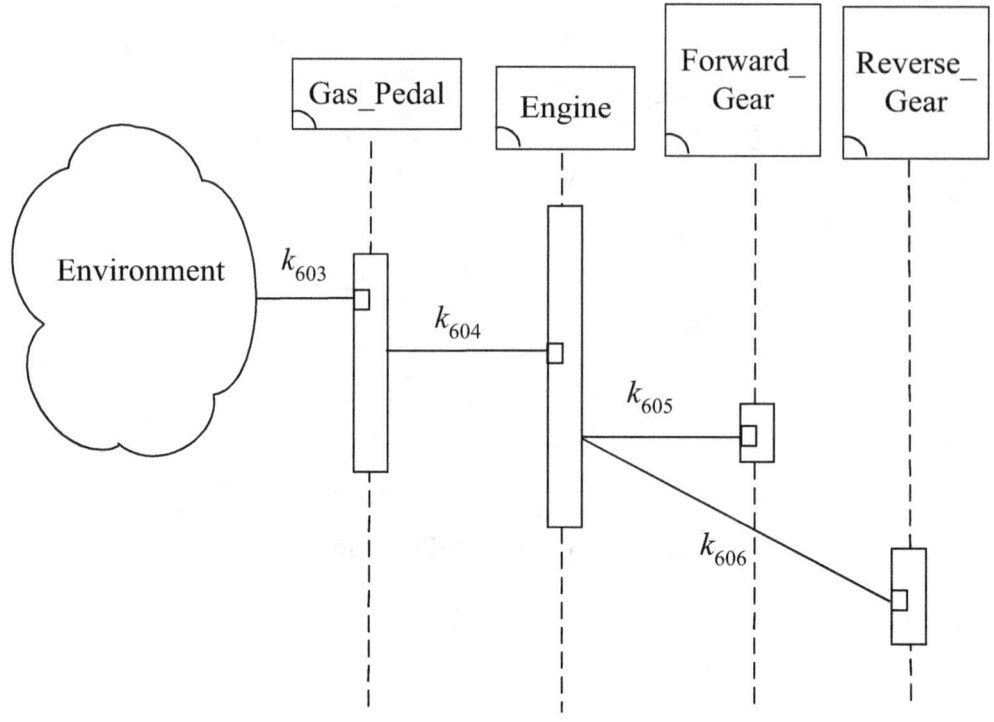

Figure 20-5   Third IFD of the *Car*

20-2 Prefixes of the Car

The t_{601} (channel-based called action with no conditions and a variable setting) prefix is defined as "<*Transmission*, CALLED, k_{601}> Gear:="Forward"", as shown in Figure 20-6.

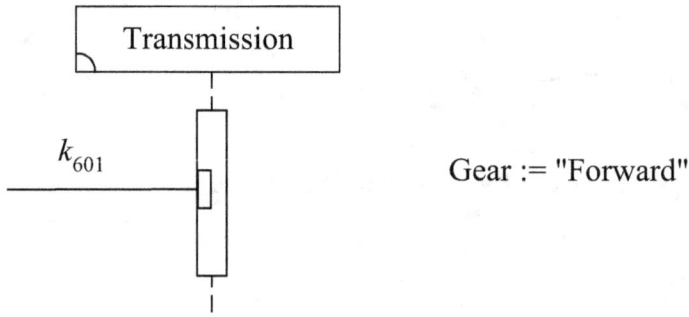

Figure 20-6   Prefix of t_{601}

The t_{602} (channel-based called action with no conditions and a variable setting) prefix is defined as "<*Transmission*, CALLED, k_{602}> Gear:="Reverse"", as shown in Figure 20-7.

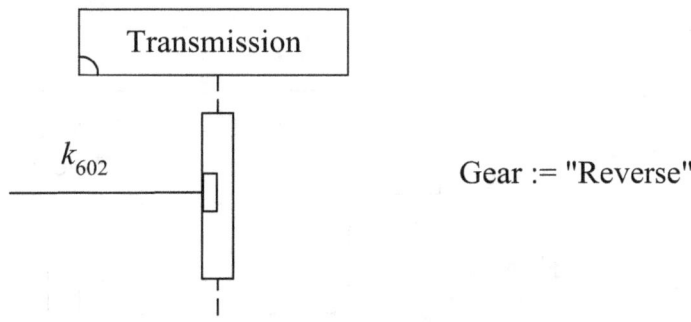

Figure 20-7   Prefix of t_{602}

The t_{603} (channel-based called action with no conditions and no variable settings) prefix is defined as <*Gas_Pedal*, CALLED, k_{603}> as shown in Figure 20-8.

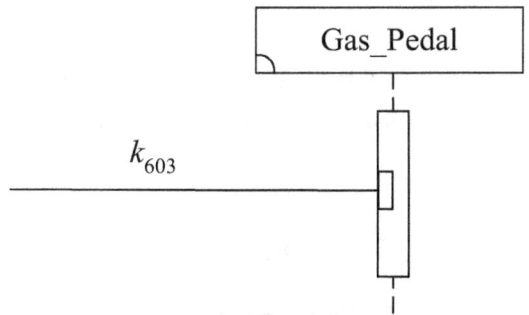

Figure 20-8   Prefix of t_{603}

The t_{604} (channel-based interaction with no conditions and no variable settings) prefix is defined as "<*Gas_Pedal*, k_{604}, *Engine*>", as shown in Figure 20-9.

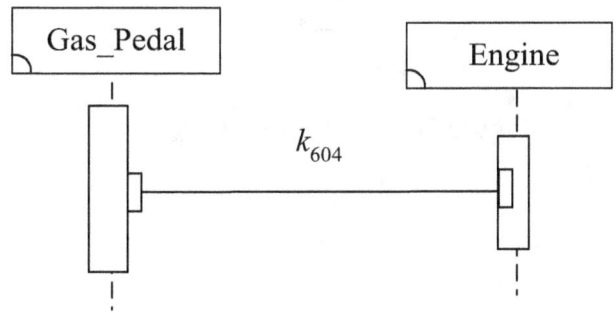

Figure 20-9   Prefix of t_{604}

The t_{605} (channel-based interaction with a condition and no variable settings) prefix is defined as "**if** Gear = "Forward" **then** <*Engine*, k_{605}, *Forward_Gear*>" as shown in Figure 20-10.

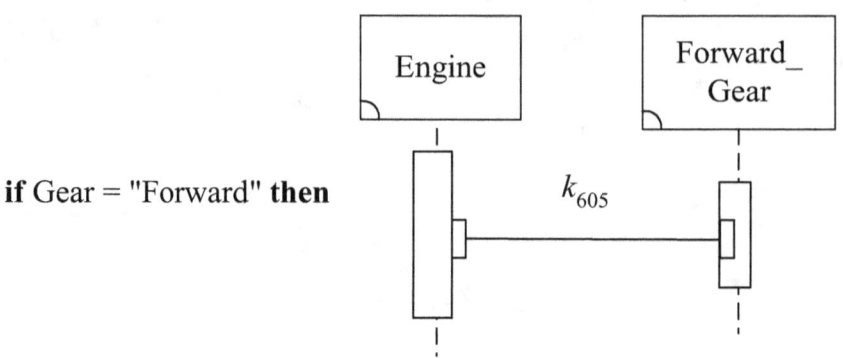

Figure 20-10   Prefix of t_{605}

The t_{606} (channel-based interaction under a certain condition) prefix is defined as "**if** Gear = "Reverse" **then** <*Engine*, k_{606}, *Reverse_Gear*>" as shown in Figure 20-11.

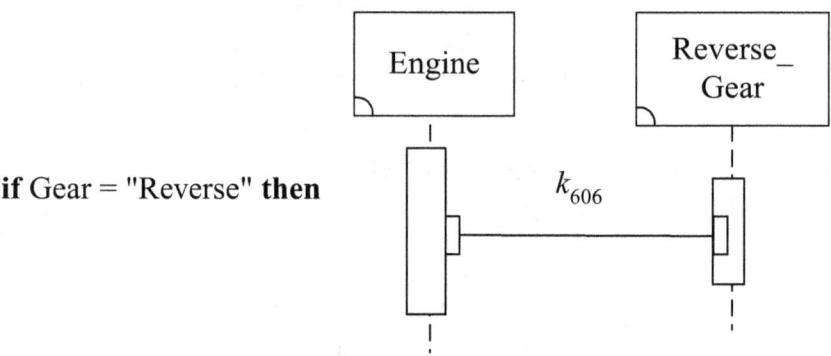

Figure 20-11   Prefix of t_{606}

Figure 20-12 shows all channel formulas of the channel-based single-queue SBC process of the *Car*.

Entity name	Channel Formula
k_{601}	Push_Gear_Forward
k_{602}	Push_Gear_Reverse
k_{603}	Depress_Gas_Pedal
k_{604}	Fuel_Supply
k_{605}	Forward_Gear_Rotate
k_{606}	Reverse_Gear_Rotate

Figure 20-12　Channel Formulas of the *Car*'s Process

Figure 20-13 shows the primitive data type specification of the *Gear* variable.

Parameter	Data Type	Instances
Gear	Enumerated	Forward, Reverse

Figure 20-13　Primitive Data Type Specification

20-3 Process of the Car

The following transition graph shows, in Figure 20-14, the semantics of A_{601}'s channel-based single-queue SBC process.

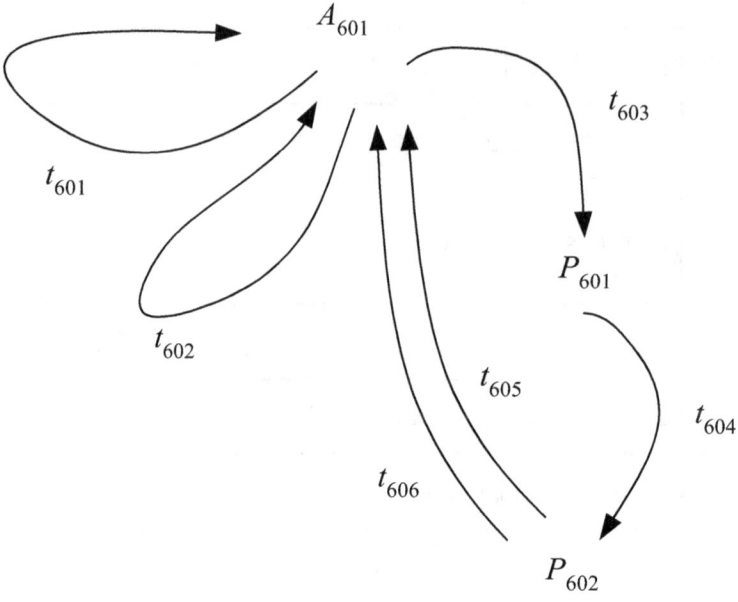

Figure 20-14   Transition graph of the *Car*'s Process

In the transition graph of the A_{601}'s channel-based single-queue SBC process, processes A_{601}, P_{601} and P_{602} are defined as in Figure 20-15.

$$A_{601} \stackrel{\text{def}}{=} t_{601} \bullet A_{601} + t_{602} \bullet A_{601} + t_{603} \bullet P_{601}$$

$$P_{601} \stackrel{\text{def}}{=} t_{604} \bullet P_{602}$$

$$P_{602} \stackrel{\text{def}}{=} t_{605} \bullet A_{601} + t_{606} \bullet A_{601}$$

Figure 20-15   Definition of Processes A_{601}, P_{601}, and P_{602}

APPENDIX A: Language Constructs of Channel-Based Single-Queue SBC Process Algebra

(1) E ::= "**fix**(" <Process_Variable> "="<IFD>{"+" <IFD>} ")"

(2) <IFD> ::= <Type_1_Expression>
    {"●" <Type_1_Or_2_Expression>} " ● " <Process_Variable>

(3) <Type_1_Or_2_Expression> ::= <Type_1_Expression>

    | <Type_2_Expression>

(4) <Type_1_Expression> ::= <Called-Action>[variable settings]

    | $\sum_{j \in I}$ **if** $cond_j$ **then** <Called-Action$_j$>[variable settings$_j$]

(5) <Type_2_Expression> ::= <Interaction>[variable settings]

    | $\sum_{j \in I}$ **if** $cond_j$ **then** <Interaction$_j$>[variable settings$_j$]

APPENDIX B: Transitional Semantics of Channel-Based Single-Queue SBC Process Algebra

Prefix $\quad \dfrac{\phantom{t \bullet E \xrightarrow{t} E}}{t \bullet E \xrightarrow{t} E}$

Sum$_j$ $\quad \dfrac{E_j \xrightarrow{t} E'_j}{\sum_{i \in I} E_i \xrightarrow{t} E'_j} \, (j \in I)$

Recursion $\quad \dfrac{\mathbf{fix}(X=z\{\mathbf{fix}(X=z)/X\}) \xrightarrow{t} E'}{\mathbf{fix}(X=z) \xrightarrow{t} E'}$

Constant $\quad \dfrac{P \xrightarrow{t} P'}{A \xrightarrow{t} P'} \, (A \stackrel{\text{def}}{=} P)$

BIBLIOGRAPHY

[Acko68] Ackoff, R., "Toward a System of Systems Concepts," *Modern Systems Research for the Behavioral Scientist: A Sourcebook*, Aldine Publishing Company, 1968.

[Berg87] Bergstra, J. A. et al., "ACPτ: A Universal Axiom System for Process Specification," *CWI Quarterly* 15, 1987, pp. 3-23.

[Burd10] Burd, S. D., *Systems Architecture*, 6th Edition, Cengage Learning, 2010.

[Chao14a] Chao, W. S., *Systems Thingking 2.0: Architectural Thinking Using the SBC Architecture Description Language*, CreateSpace Independent Publishing Platform, 2014.

[Chao14b] Chao, W. S., *General Systems Theory 2.0: General Architectural Theory Using the SBC Architecture*, CreateSpace Independent Publishing Platform, 2014.

[Chao14c] Chao, W. S., *Systems Modeling and Architecting: Structure-Behavior Coalescence for Systems Architecture*, CreateSpace Independent Publishing Platform, 2014.

[Chao15a] Chao, W. S., *Variants of Interaction Flow Diagrams: The Structure-Behavior Coalescence Approach*, CreateSpace Independent Publishing Platform, 2015.

[Chao15b] Chao, W. S., *A Process Algebra For Systems Architecture: The Structure-Behavior Coalescence Approach*, CreateSpace Independent Publishing Platform, 2015.

[Chao15c] Chao, W. S., *An Observation Congruence Model For Systems Architecture: The Structure-Behavior Coalescence Approach*, CreateSpace Independent

Publishing Platform, 2015.

[Chao15d] Chao, W. S., *Variants of SBC Process Algebra: The Structure-Behavior Coalescence Approach*, CreateSpace Independent Publishing Platform, 2015.

[Chao15e] Chao, W. S., *Single-Queue SBC Process Algebra For Systems Architecture: The Structure-Behavior Coalescence Approach*, CreateSpace Independent Publishing Platform, 2015.

[Chao15f] Chao, W. S., *Multi-Queue SBC Process Algebra For Systems Architecture: The Structure-Behavior Coalescence Approach*, CreateSpace Independent Publishing Platform, 2015.

[Chao15g] Chao, W. S., *Single-Queue SBC Observation Congruence Model For Systems Architecture: The Structure-Behavior Coalescence Approach*, CreateSpace Independent Publishing Platform, 2015.

[Chao15h] Chao, W. S., *Multi-Queue SBC Observation Congruence Model For Systems Architecture: The Structure-Behavior Coalescence Approach*, CreateSpace Independent Publishing Platform, 2015.

[Chao16a] Chao, W. S., *System: Contemporary Concept, Definition, and Language*, CreateSpace Independent Publishing Platform, 2016.

[Chao16b] Chao, W. S., *Systems Architecture of Electronic Toll Collection Cloud Applications and Services IoT System*, CreateSpace Independent Publishing Platform, 2016.

[Chao16c] Chao, W. S., *Systems Architecture of Smart Healthcare Cloud Applications and Services IoT System*, CreateSpace Independent Publishing Platform, 2016.

[Chao16d] Chao, W. S., *Systems Architecture of Smart Healthcare Cloud Applications and Services IoT System*, CreateSpace Independent Publishing

Platform, 2016.

[Chao16e] Chao, W. S., *Systems Architecture of Ridesharing Sharing Economy Cloud Applications and Services IoT System*, CreateSpace Independent Publishing Platform, 2016.

[Chao16f] Chao, W. S., *Systems Architecture of Handy Helper Sharing Economy Cloud Applications and Services IoT System*, CreateSpace Independent Publishing Platform, 2016.

[Chao16g] Chao, W. S., *Systems Architecture of Vacation Rental Cleaning Sharing Economy Cloud Applications and Services IoT System*, CreateSpace Independent Publishing Platform, 2016.

[Chec99] Checkland, P., *Systems Thinking, Systems Practice: Includes a 30-Year Retrospective*, 1st Edition, Wiley, 1999.

[Craw15] Crawley, P. et al., *System Architecture: Strategy and Product Development for Complex Systems*, Prentice Hall, 2015.

[Dam06] Dam, S., *DoD Architecture Framework: A Guide to Applying System Engineering to Develop Integrated Executable Architectures*, BookSurge Publishing, 2006.

[Date03] Date, C. J., *An Introduction to Database Systems*, 8th Edition, Addison Wesley, 2003.

[Elma10] Elmasri, R., *Fundamentals of Database Systems*, 6th Edition, Addison Wesley, 2010.

[Frie11] Friedenthal, S., et al., *A Practical Guide to SysML: The Systems Modeling Language*, Morgan Kaufmann, 2nd Edition, 2011.

[Ghar11] Gharajedaghi, J., *Systems Thinking: Managing Chaos and Complexity: A Platform for Designing Business Architecture*, Morgan Kaufmann, 2011.

[Hoar85] Hoare, C. A. R., *Communicating Sequential Processes*, Prentice-Hall, 1985.

[Maie09] Maier, M. W., *The Art of Systems Architecting*, 3rd Edition, CRC Press, 2009.

[Mead08] Meadows, D. H., *Thinking in Systems: A Primer*, Chelsea Green Publishing, 2008.

[Miln89] Milner, R., *Communication and Concurrency*, Prentice-Hall, 1989.

[Miln99] Milner, R., *Communicating and Mobile Systems: the π-Calculus*, 1st Edition, Cambridge University Press, 1999.

[O'Rou03] O'Rourke, C. et al, *Enterprise Architecture Using the Zachman Framework*, 1st Edition, Course Technology, 2003.

[Putm00] Putman, J. R. et al., *Architecting with RM-ODP*, Prentice-Hall, 2000.

[Rayn09] Raynard, B., *TOGAF The Open Group Architecture Framework 100 Success Secrets*, Emereo Pty Ltd, 2009.

[Roza11] Rozanski, N. et al., *Software Systems Architecture: Working With Stakeholders Using Viewpoints and Perspectives*, 2nd Edition, Addison-Wesley Professional, 2011.

[Scho10] Scholl, C., *Functional Decomposition with Applications to FPGA Synthesis*, Springer, 2010.

[Toga08] The Open Group, *TOGAF Version 9 - A Manual (TOGAF Series)*, 9th Edition, Van Haren Publishing, 2008.

INDEX

A

ACP. *See* algebra of communicating processes

action

    called action, 33

    calling action, 33

agent

    callee, 33

    caller, 32

algebra of communicating processes, 23

B

building block. *See* component

C

calculus of communicating systems, 23

callee, 33

caller, 32

CCS. *See* calculus of communicating systems

channel formula, 31

channel-based action, 36

channel-based infinite-queue SBC process algebra, 24

channel-based interaction, 32

channel-based multi-queue SBC process algebra, 24

channel-based port, 35

channel-based single-queue SBC process algebra, 24

C-I-SBC-PA. *See* channel-based infinite-queue SBC process algebra

C-M-SBC-PA. *See* channel-based multi-queue SBC process algebra

communicating sequential processes, 23

communication. *See* interaction

component, 40

CSP. *See* communicating sequential processes

C-S-SBC-PA. *See* channel-based single-queue SBC process algebra

E

entity. *See* component

external environment, 40

F

functional decomposition, 20, 22

G

generalized SBC process algebra, 23

graphical user interface, 62, 77

G-SBC-PA. *See* generalized SBC process algebra

GUI. *See* graphical user interface

H

handshake. *See* interaction

I

IFD. *See* interaction flow diagram

interaction, 22, 32

interaction flow diagram, 45, 46, 56, 57, 67, 82, 94, 107, 116, 117, 123, 124, 132, 133, 141, 142, 151, 152, 161, 162, 171, 172, 183, 184

internal interaction, 35

L

labelled transition system, 49

LTS. *See* labelled transition system

N

non-aggregated system. *See* component

O

object. *See* component

O-I-SBC-PA. *See* operation-based infinite-queue SBC process algebra

O-M-SBC-PA. *See* operation-based multi-queue SBC process algebra

operation-based infinite-queue SBC process algebra, 24

operation-based multi-queue SBC process algebra, 24

operation-based single-queue SBC process algebra, 24

O-S-SBC-PA. *See* operation-based single-queue SBC process algebra

P

parallel composition, 25

part. *See* component

port

    called port, 33

    calling port, 32

prefix, 50

process algebra, 9, 23

    algebra of communicating processes, 23

    calculus of communicating systems, 23

    channel-based infinite-queue SBC process algebra, 24

    channel-based multi-queue SBC process algebra, 24

    channel-based single-queue SBC process algebra, 24

    communicating sequential processes, 23

    generalized SBC process algebra, 23

    operation-based infinite-queue SBC process algebra, 24

    operation-based multi-queue SBC process algebra, 24

    operation-based single-queue SBC process algebra, 24

R

recursion, 26

replication, 26

S

SBC. *See* structure-behavior coalescence

sequentialization, 25

structural decomposition, 20, 22

structure-behavior coalescence, 39, 40

summation, 25

system, 15, 16

    abstract system, 16

    boundary, 17

    concrete system, 15

    environment, 17

    isolated system, 19

    notional system, 16

    open system, 18

    physical system, 15

    real system, 15

    virtual system, 16

systems, 39, 40

systems behavior, 39, 40

systems modeling, 9, 15, 24

    systems modeling 1.0, 19

    systems modeling 2.0, 21

systems structure, 39, 40

T

transition graph, 59, 75, 88, 100, 113, 120, 129, 136, 147, 157, 166, 178, 188

V

variable setting, 43, 49, 50, 51, 172, 184, 185

www.ingramcontent.com/pod-product-compliance
Lightning Source LLC
Chambersburg PA
CBHW081145180526
45170CB00006B/1940